T0205555

Studies in Systems, Decision and Control

Volume 131

Series editor

Janusz Kacprzyk, Polish Academy of Sciences, Warsaw, Poland
e-mail: kacprzyk@ibspan.waw.pl

The series "Studies in Systems, Decision and Control" (SSDC) covers both new developments and advances, as well as the state of the art, in the various areas of broadly perceived systems, decision making and control- quickly, up to date and with a high quality. The intent is to cover the theory, applications, and perspectives on the state of the art and future developments relevant to systems, decision making, control, complex processes and related areas, as embedded in the fields of engineering, computer science, physics, economics, social and life sciences, as well as the paradigms and methodologies behind them. The series contains monographs, textbooks, lecture notes and edited volumes in systems, decision making and control spanning the areas of Cyber-Physical Systems, Autonomous Systems, Sensor Networks, Control Systems, Energy Systems, Automotive Systems, Biological Systems, Vehicular Networking and Connected Vehicles, Aerospace Systems, Automation, Manufacturing, Smart Grids, Nonlinear Systems, Power Systems, Robotics, Social Systems, Economic Systems and other. Of particular value to both the contributors and the readership are the short publication timeframe and the world-wide distribution and exposure which enable both a wide and rapid dissemination of research output.

More information about this series at http://www.springer.com/series/13304

Paolo Massimo Buscema
Giulia Massini · Marco Breda
Weldon A. Lodwick · Francis Newman
Masoud Asadi-Zeydabadi

Artificial Adaptive Systems Using Auto Contractive Maps

Theory, Applications and Extensions

 Springer

Paolo Massimo Buscema
Semeion Research Center of Sciences
   of Communication
Rome
Italy

Giulia Massini
Semeion Research Center of Sciences
   of Communication
Rome
Italy

Marco Breda
Semeion Research Center of Sciences
   of Communication
Rome
Italy

Weldon A. Lodwick
Department of Mathematical
   and Statistical Sciences
University of Colorado Denver
Denver, CO
USA

Francis Newman
Department of Radiation Oncology,
   School of Medicine
University of Colorado Denver
Denver, CO
USA

Masoud Asadi-Zeydabadi
Physics Department
University of Colorado Denver
Denver, CO
USA

ISSN 2198-4182   ISSN 2198-4190   (electronic)
Studies in Systems, Decision and Control
ISBN 978-3-030-09135-4   ISBN 978-3-319-75049-1   (eBook)
https://doi.org/10.1007/978-3-319-75049-1

Printed on acid-free paper

This Springer imprint is published by the registered company Springer International Publishing AG part of Springer Nature
The registered company address is: Gewerbestrasse 11, 6330 Cham, Switzerland

Contents

Chapter 1
An Introduction

> *Nearly all persons look not at theory from the standpoint of established facts, you know, but at established facts from the standpoint of theory; they cannot get beyond an assumed conceptual net they have accepted, ...*
> Einstein (letter to Schroedinger, 8 August 1935, in Albrecht Fölsing, *Albert Einstein, Eine Biographie*, page 780, Suhrkamp Verlag, 1993)

Abstract Auto-Contractive Maps (Auto-CM) is a newer approach to artificial adaptive systems (AASs). In turn, AASs encompass the subject of artificial neural networks (ANNs). This chapter is an introduction to AASs.

1.1 Introduction

Artificial adaptive systems (AASs) are data driven systems. This means that AASs try to elicit, extract, from data the underlying model of cause/effect. These methods stand in counter distinction to approaches that impose a model on the data as a means of determining the relationship (cause and effect for example) among the data elements. Informally speaking, for this monograph, we consider *data* as being what is known about the system or problem, its input elements. The philosophy of AAS can be described as and compared to natural/human language. Some parallels are striking. The artificial sciences try to create models of reality, but how well they approximate the "world" determines their effectiveness and usefulness. Human languages, in a similar manner, try to approximate the reality of the subject at hand. There is another similarity between artificial adaptive systems and human languages and that is that both undergo dynamic changes to more closely resemble the "reality" of the entities of interest.

Auto Contractive Maps that is the main topic of this book is an artificial adaptive system, an artificial neural network, that adheres to the approach that Einstein mentions in his letter to Schroedinger. Auto-CM extracts the model from the data as will be apparent in our subsequent presentation beginning with Chap. 3 of this monograph.

© Springer International Publishing AG, part of Springer Nature 2018   1
P. M. Buscema et al., *Artificial Adaptive Systems Using Auto Contractive Maps*, Studies in Systems, Decision and Control 131, https://doi.org/10.1007/978-3-319-75049-1_1

Artificial adaptive systems include much of artificial intelligence and artificial neural networks (ANNs). This monograph restricts itself to one particular type of unsupervised ANN called Auto-Contractive Map (Auto-CM) and its supervised version called K-Auto CM. The Auto-CM approach is a newer type of ANN that is powerful enough as a stand-alone analytical tool. Auto-CM takes a different approach to ANNs in that its analytical method is more akin to fixed-point algorithms where, in our case, the values at the nodes converge to zero distributing their input dataset values to the weights. The final output of Auto-CM indicates a relationship among the variables of the dataset and these are found in the weights. Traditional ANNs attempt to stabilize the weights via some error minimization techniques. The weights of Auto-CM are also stabilized in the sense that the algorithm stops when all the node values are essentially zero but this occurs because a fixed-point has been reached rather than a minimum of an error function, for example, has been attained.

We not only present Auto-CM but show how to couple Auto-CM to two associated graphical visualization components, the minimal spanning tree (MST) and maximal regular graph (MRG) in order to graphically (in the graph theoretic sense) depict the relationships in the dataset. In addition to these two visualization tools and their interpretation of the underlying relationship in the data, we present methods to transform the output of the neural network (our Auto-CM output) into new datasets that allow for deeper interpretation of the data relationships akin to *deep learning* methods. In particular, the original dataset may be transformed via Auto-CM into new databases that have a richer set of relationships among variables and records; for example, fuzzy relationships and relationships that "collapse" variables and records. Our point of view is that the universe is not random and at the same time it loves to make us work hard to understand her. Thus, ferreting out the "patterns" that are in data is not easy. The ferreting out of patterns is what Auto-CM attempts to do. Auto-CM attempts to uncover nature's secrets.

What we present targets users, researchers and students of ANNs who wish to see what a more ample approach to ANNs might look like. Only basic notions of mathematics are assumed as well as an understanding of neural networks, though we derive key features. Our level of presentation is at the upper level undergraduate to lower level graduate student in computer science, mathematics, physics or engineering.

This monograph continues this section with a general discussion of where Auto-CM fits within the field of artificial intelligence. Chapter 2 has an overview of the standard ANNs in order to clearly contrast the features of Auto-CM. Chapter 3 contains the core of our monograph, Auto-CM, an unsupervised type of ANN. Chapter 4 presents the graphical component of Auto-CM. Chapter 5 looks at how the Auto-CM output can be further transformed into new datasets that contain a wider view of the relationships that exist among data elements. Chapter 6 presents a supervised version of Auto-CM called K-Contractive Map. Chapter 7 compares Auto-CM with various ANNs. We end this monograph with Chap. 8 which has more advanced notions of how to add dynamic changes in time and spaces, to Auto-CM and hence ANNs.

1.2 What Are Artificial Adaptive Systems

Our introduction begins with the differences between deterministic/stochastic and adaptive approaches. Deterministic/stochastic systems are often called complicated systems whereas adaptive systems are often called complex systems.

A system is a connected region of space-time, divided into components, whose local and parallel interactions determine the functioning of the system itself. A system, whose operation (interactions between its components) dynamically changes the structure and/or the status of its members, while maintaining its space-time cohesion, is what we call a *complex system*. A human cell that becomes a human or a social organism, such as a group of scholars that becomes an academic department, is a complex system.

A system whose complete permutation of all of its components allows the definition of all possible states of the system is a *complicated system*, the more complicated the more numerous are its possible states. For example, a jumbo jet aircraft, with all of its 200,000 small components, is complicated but is not a complex system. A movie is a complicated system. A complicated system is a system whose operation does not generate new information since every possible trajectory is defined a priori.

A complicated system, which works over time, ages since its performance remains static. In a complex system the global behavior of the system is not inferred from the simple sum of the behaviors of all of its components, or by linear interpolation techniques. The overall behavior of a complex system is behavior that emerges from the operation of the system *over time*. This process is therefore highly non-linear, which in many cases is not formalized by equations in closed form. A complex system is one whose operation generates new information over time, and over time changes the structure and/or the state of its components, which we call an *adaptive system*.

An artificial system is a model of a part of reality that is encoded in mathematical relationships and implemented on a computer. Our view of an artificial adaptive system is depicted in Fig. 1.1 below where evolutionary programming and ANNs fall under the umbrella of AASs. Most of the features of Fig. 1.1 will be discussed in Chap. 2.

The dynamics of an adaptive system necessitates a mathematics that does not impose linear assumptions on the data. The mathematics that is needed works on data using a "Socratic" style. Ideally, it works in a way in which the overall behavior of the system emerges spontaneously by the local interaction of its components represented by *data and equations*. This is the method of a "natural" operation, from the identification of the targeted problem up to the interpretation of the results. To achieve this, we look at problems as research questions on which we undertake experiments via computer algorithms, simulations. This is the method of "Natural Computation" (Bottom-Up) and it is the best way for the study of natural adaptive systems (Buscema 2010). AASs are the mathematical expression of the Natural Computation method. Their purpose is to bring out the overall operation of a natural adaptive system, whose data represent discrete portions of the functioning system.

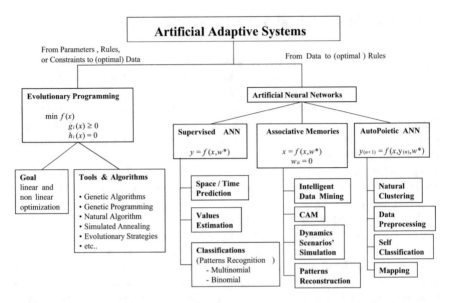

Fig. 1.1 Artificial adaptive systems

We will use the expression "cultural system" to mean a natural system whose operation is filtered in a non-linear interaction between human brains. In natural systems and, therefore, also cultural ones, there is hidden information that lies within the data and via AASs can be disclosed and become perceptible among the relationships discovered in the data, and its information which is more visible and easily measureable. The hidden information reveals the "secret plan" that the natural adaptive system is going to pursue.

Artificial Adaptive Systems look for traces of the hidden agenda of natural adaptive systems. The range of these applications is also related to Intelligent Data Mining ([1] and see Fig. 1.2).

AASs emerge by subjecting the system to experiments. An "experiment" is the design, execution, and completion of data processing from the initial treatment of a selected or a complete set of available data including the interpretation of the local results. By "simulation" we mean the process that starts from the application of the different artificial adaptive algorithms to the data, up to the generation of results (see Fig. 1.3). The category "research" in Fig. 1.3 may be considered to be related to the same problem as different facets or they may be viewed as independent.

Figure 1.3 illustrates what we consider to be a robust research protocol and it is how we validated all our algorithms and examples. An AAS can be thought of being implemented as follows. We assume we have identified the problem, which means we have collected significant and non-trivial questions and facts about a specific natural and/or cultural system of interest. It also means that we have speculated about the possible operation of such a system, that is, we have formed hypotheses which can be methodologically verified in the following steps.

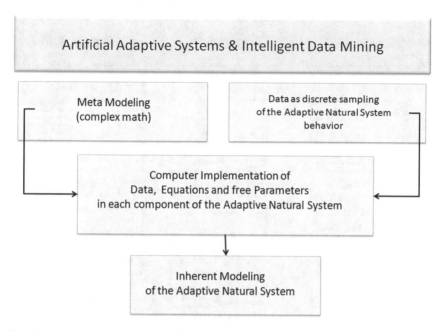

Fig. 1.2 The main components of Intelligent Data Mining schematized in order to understand natural adaptive systems

1. **The Data**. The study's questions and/or hypotheses need to be translated into structured input entities or data. That is, data is what we know about the system. Such data must be statistically representative of the process that we want to examine. The data must be collected with appropriate statistical criteria and/or must be generated by special experiments/trials. In all cases, the data that is available will be the only input for the next steps of the research process. All of the research has to be completely data driven.

2. **Data Pre-processing**. The data available for research should be organized and structured with appropriate algorithms, in order to create a database presenting all the information in an explicit format. Consequently, adaptive algorithms that will be used later, will facilitate the discovery of distributed and/or hidden information and structures in the database.

3. **Adaptive Algorithms**. Choose the AAS algorithms that are appropriate for the research and arrange them in sequential and/or parallel structures. The choice and the organization of the algorithms in a framework define research design.

4. **Blind Validation Methods**. The individual algorithms and the research design must be validated in a blind fashion, splitting the data numerous times into training and testing data subsets. There are various protocols available to implement the research's validation phase, and the most appropriate protocol has to be chosen for the assigned research. In addition, a suitable cost function that measures the

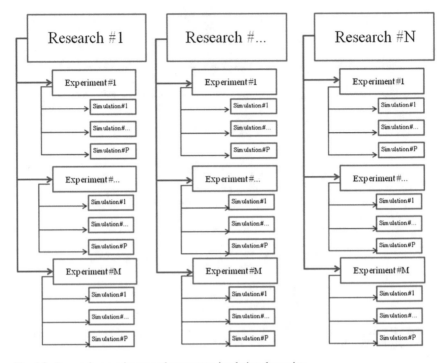

Fig. 1.3 Research, experiment and computer simulation dynamics

results of the different simulations has to be selected, according to statistical criteria.

5. **Results Interpretation**. This phase is very delicate. We need to interpret the different outputs of the experiment in relation to the questions and assumptions we made in the initial phase of the research when the experimental simulation proves itself to be significantly correct. This phase should lead to new hypotheses and/or questions that allow the experimental cycle to start from the beginning. These new hypotheses might be based on comparisons with known results or on predictions that were made requiring re-evaluation of the assumptions.

6. **Storage of New Information**. The storage of the results obtained, and their interpretation, is new knowledge that was not present in the initial phase of the research. This knowledge should be structured and organized in a new database, upgradeable over time, and can possibly be the object of further experiments.

The content of any research should not to be constrained a priori for two reasons. First, data must be the only source of knowledge of the targeted problem for the adaptive algorithms. In other words, any constraints and/or extra knowledge about the problem, has to be expressed in the form of quantitative and/or qualitative data. Second, artificial adaptive systems are equipped with a specific capacity to learn the complex dynamics of any process only from the data. This data have to be provided in a statistically meaningful manner. AASs, therefore, are algorithms that do not follow specific a priori rules. They do not optimize the relationships between linear data. Artificial adaptive systems *learn* directly from the data and they *evolve* in order to define the mathematical function that interpolates the assigned data in an optimal or near-optimal way. In this sense, artificial adaptive systems can be called "Meta-Models", because they learn the *intrinsic model* from the data.

1.3 The Components

The components of a research problem involving AASs can be briefly described by three components (see Fig. 1.4):

1. Databases;
2. Artificial Adaptive System Algorithms;
3. Validation Protocols.

Some recommended strategies in carrying out research from the point of view of an AAS process are:

1. Carry out many different experimentations that fail, in order to "grow up"; failure is a necessary condition and an opportunity to be embraced for needed development, innovation and changes;
2. Learn to consider and to look at any process from the same point of view that nature has adopted in order to generate that process;
3. Pay attention to the relationships among the objects, because relationships among objects are perceived before the objects, for example, when you see two points

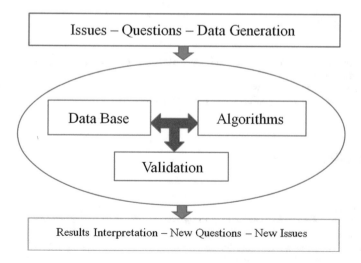

Fig. 1.4 The research process using artificial adaptive system methodology

in the space, your gaze is the relationship *between* the two points, without your gaze, the two points may not be linked together;

4. "Listen" to the weak links among things because the weak links often explain the strong(er) ones, for example, there seems to be a weak link between the head and the feet. But many headaches depend upon the way in which we move our feet and our posture;

5. Use mathematics to transform any random "scribble" into a possible pattern, because nature may in certain circumstances try to ignore noise. An addative inverse relationship between variance and entropy may exist in certain circumstances [2].

In short, what we invite the reader to consider and to explore in the following pages is achieved by using only three aspects:

a. Data of the process targeted to be investigated, the "problem";
b. The use of AAS algorithms in computer simulations of the problem;
c. Blind validation tests.

Another aspect that helps during experiments or research trials is the capability to be completely *coherent* inside every experiment and completely *incoherent* going from one experiment to another. That is, to test a theory one needs to be coherent and analytic within each experiment, then "jump" to a completely different experiment in order to test the same theory.

We end this section by reiterating that the topic of interest to this monograph is the unsupervised ANN called Auto-CM and its supervised version together with their connections to graphs and database transformations. Therefore, we will, in the next chapter, briefly review ANNs.

Specific References

1. Buscema, M., William J. Tastle. 2013. Data Mining Applications Using Artificial Adaptive Systems, Springer.
2. Swanson, R., and S.M. Swanson. 1993. The Effect of Noise on Entropy. *Acta Crystallographica. Section D, Biological Crystallography* 49 (Pt 1): 182–185.

General References

Anderson J.A., and E. Rosenfeld (eds.). 1988. *Neurocomputing Foundations of Research*, Cambridge, MA: The MIT Press.
Arbib, M.A. (ed.). 1995. *The Handbook of Brain Theory and Neural Networks, A Bradford Book*. Cambridge, Massachusetts, London, England: The MIT Press.
Bengio, Y. 2009. Learning Deep Architectures for AI. *Machine Learning* 2 (1): 1–127.
Buscema, M. 1998. Artificial Neural Networks and Complex Social Systems. In *Substance Use & Misuse*, vol. 33(1): Theory, vol. 33(2): Models, vol. 33(3): Applications, New York: Marcel Dekker.

Buscema, M. 1999. Semeion Group. In *Reti Neurali Artificiali e Sistemi Sociali Complessi*, vol. I: Teoria e Modelli, vol. II: Applicazioni, Franco Angeli, Milano.

Carpenter, G.A., and S. Grossberg. 1991. *Pattern Recognition by Self-Organizing Neural Network*. Cambridge, MA: MIT Press.

Grossberg, S. 1978. How Does the Brain Build a Cognitive Code? *Psychological Rewiew* 87.

Hinton, G.E., S. Osindero, and Y-W. Teh. 2006. A Fast Learning Algorithm for Deep Belief Nets. *Neural Computation* 18 (7): 1527–1554 (MIT Press Journal).

Hinton, G.E., and R.R. Salakhutdinov. 2006. Reducing the Dimensionality of Data with Neural Networks. *Science* 313: 504–507.

Hopfield, J.J. 1987. Neural Networks and Physical Systems with Emergent Collective Computational Abilities. In *Proceedings of the National Academy of Sciences* 79, in Anderson 1988.

Hopfield, J.J. 1984. Neurons with Graded Response have Collective Computational Properties Like Those of Two-State Neurons. In *Proceedings of the National Academy of Sciences USA, Bioscience*, 81.

Hopfield, J.J., and D.W. Tank. 1985. Neural Computation of Decisions in Optimization Problems. *Biological Cybernetics* 52.

Hopfield, J.J., and D.W. Tank. 1986. Computing with Neural Circuits: A Model. *Articles Science* 233: 8.

Kohonen T. 1972. Correlation Matrix Memories. *IEEE Transactions on Computers C-21*, in Anderson 1988.

Larochelle, H., and Y. Bengio. 2008. Classification using Discriminative Restricted Boltzmann Machines. In *Proceedings of the 25-th International Conference on Machine Learning*, Helsinki, Finland.

Le, Q.V., M.C. Ranzato, R. Monga, M. Devin, K. Chen, G.S. Corrado, J. Dean, and A.Y. Ng. 2012. Building High-level Features Using Large Scale Unsupervised Learning, In *Proceedings of the 29-th International Conference on Machine Learning*, Edinburgh, Scotland, UK.

Maturana, H., and F. Varela. 1980. *Autopoiesis and Cognition: The Realization of the Living*, Springer Science.

Minsky M. 1954. Neural Nets and the Brain-Model Problem, *Doctoral dissertation*, Princeton University.

Minsky, M., and S. Papert. 1988. *Perceptrons, extended ed*. Cambridge, MA: The MIT Press.

Raiko, T., H. Valpola, and Y. LeCun. 2012. Deep Learning Made Easier by Linear Transformations in Perceptrons. In *Proceedings of the 15th International Conference on Artificial Intelligence and Statistics (AISTATS) 2012*, vol. XX of JMLR:W and CP XXLa, Canary Islands: Palma.

Raina, R., A. Madhavan, and A.Y. Ng. 2009. Large-Scale Deep Unsupervised Learning using Graphics Processors. In *Proceedings of the 26th International Conference on Machine Learning*, Montreal, Canada.

Reetz, B. 1993. Greedy Solution to the Travelling Sales Person Problem. *ATD 2*.

Rosenblatt, F. 1962. *Principles of Neurodynamics*. N.Y.: Spartan.

Rumelhart D.E., J. L. Mcclelland (eds.). 1986. *Parallel Distributed Processing, vol. 1 Foundations, Explorations in the Microstructure of Cognition, vol. 2 Psychological and Biological Models*. Cambridge, MA, London, England: The MIT Press.

Werbos, P. 1974. Beyond Regression: New Tools for Prediction and Analysis in Behavioral Sciences, *PhD thesis*, Cambridge, MA: Harvard.

Chapter 2
Artificial Neural Networks

Abstract Artificial Adaptive Systems include Artificial Neural Networks (ANNs or simply neural networks as they are commonly known). The philosophy of neural networks is to extract from data the underlying model that relates this data as an input/output (domain/range) pair. This is quite different from the way most mathematical modeling processes operate. Most mathematical modeling processes normally impose on the given data a model from which the input to output relationship is obtained. For example, a linear model that is a "best fit" in some sense, that relates the input to the output is such a model. What is imposed on the data by artificial neural networks is an a priori architecture rather than an a priori model. From the architecture, a model is extracted. It is clear, from any process that seeks to relate input to output (domain to range), requires a representation of the relationships among data. The advantage of imposing an architecture rather than a data model, is that it allows for the model to adapt. Fundamentally, a neural network is represented by its architecture. Thus, we look at the architecture first followed by a brief introduction of the two types of approaches for implementing the architecture—supervised and unsupervised neural networks. Recall that Auto-CM, which we discuss in Chap. 3, is an unsupervised ANN while K-CM, discussed in Chap. 6, is a supervised version of Auto-CM. However, in this chapter, we show that, in fact, supervised and unsupervised neural networks can be viewed within one framework in the case of the linear perceptron. The chapter ends with a brief look at some theoretical considerations.

2.1 Introduction

We begin with the anatomy of neural networks, its architecture. ANNs are a family of methods inspired by the human brain's learning capability. ANNs are scientifically used in three different epistemological ways:

1. To understand, the function of the brain by computer simulations;
2. To reproduce in computer algorithms, the way the brain functions in its relationship with the environment, for example, in problem solving, driving a car, and so on (human brain emulation);

© Springer International Publishing AG, part of Springer Nature 2018
P. M. Buscema et al., *Artificial Adaptive Systems Using Auto Contractive Maps*, Studies in Systems, Decision and Control 131, https://doi.org/10.1007/978-3-319-75049-1_2

3. To understand the transition from individual to collective behavior (data analysis, data mining and the research on complex systems are part of this).

Currently ANNs comprise a range of very different models, but they all share the following characteristics.

- The fundamental elements of ANNs are the *nodes*, also known as processing elements and their *connections*.
- Each node in an ANN has its own *input* through which it receives communications from the other nodes or from the environment, and its own *output*, through which it communicates with other nodes or with the environment. In addition, it has an internal function, $f(\cdot)$ which transforms its global input into an output.
- Each connection may poses an internal relationship or "force" between pairs of nodes that excite or inhibit each other. Positive values indicate excitatory connections and negative ones indicate inhibitory connections.
- Connections between nodes may change over time. This dynamic (time dimension) triggers a *learning process* throughout the entire ANN. The way the law by which the connections change in time is called the "learning equation".
  In order for the connections of the ANN to properly change, the environment must act on the ANN several times.
- When ANNs are used to process data, the data are ANN's environment. Thus, in order to process data, the data is subjected to the ANN several times.
- The overall dynamic of an ANN depends exclusively on the local interaction of its nodes. The final state of the ANN "evolves" "spontaneously" from the interaction of all of its components (nodes).
- Communications between nodes in every ANN tend to occur in *parallel*. This parallelism may be *synchronous* or *asynchronous* and each ANN may emphasize this parallelism in a different way. In synchronous ANNs, all nodes simultaneously update their state variables whereas in the asynchronous regime a random node is chosen to update and the other nodes likewise are updated in a random fashion. However, an ANN always has some form of parallelism in the activity of its nodes. From a theoretical viewpoint this parallelism does not depend on the hardware on which the ANNs are implemented.

The architecture of every ANN is composed of the following five components:

1. Type and number of *nodes* and their corresponding properties;
2. Type and number of *connections* and their corresponding location;
3. Type and number of *layers*;
4. Type of *signal flow* strategy;
5. Type of *learning strategy*.

In short, ANNs have *nodes, connections, layers, signal flow,* and *learning strategy*. These five aspects are discussed next.

The Nodes

There are three types of ANN nodes, depending on the position they occupy within the ANN.

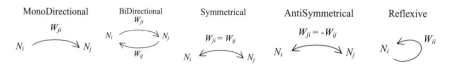

Fig. 2.1 Types of possible connections

1. *Input nodes*: the nodes that (also) receive signals from the environment outside the ANN.
2. *Output nodes*: the nodes whose signal (also) acts on the environment outside the ANN.
3. *Hidden nodes*: the nodes that receive signals only from other nodes "inside" the ANN and send their signal only to other nodes "inside" the ANN.

The number of input nodes depends on the way the ANN is intended to read the environment. The input nodes are the ANN's *sensors*. When the ANN's environment consists of data, the ANN should process the input node as a type of data *variable*. The number of output nodes depends on the way one wants the ANN to act on the environment. The output nodes are the *effectors* of the ANN. When the ANN's environment consists of data to process, the output nodes represent the variables sought or the results of the processing that occurs within the ANN. The number of hidden nodes depends on the complexity of the function one intends to map between the input nodes and the output nodes. The nodes of each ANN may be grouped into classes of nodes sharing the same properties. Normally these classes are called layers.

The Connections

There may be various types of connections: *Mono-Directional, Bi-Directional, Symmetrical, Anti-Symmetrical and Reflexive* (see Fig. 2.1).

The number of connections is proportional to the memory capabilities of the ANN. Positioning the connections may be useful as methodological preprocessing for the problem to be solved, but it is not necessary. An ANN in which the connections between nodes or between layers are not all enabled is called an ANN with dedicated connections; otherwise it is known as a maximum gradient ANN.

Each ANN has connections that can be:

- *Adaptive*: they change depending on the learning equation.
- *Fixed*: they remain at fixed values throughout the learning process.
- *Variable*: they change deterministically as other connections change.

The Layers

Various types of layers can be distinguished. Among them, we mention the three most important.

1. *Mono-Layer* ANNs: all nodes of the ANN have the same properties.
2. *Multi-Layer* ANNs: the ANN nodes are grouped in functional classes; for example, nodes that (a) share the same signal transfer functions or (b) receive the signal only from nodes of other layers and send them only to new layers.
3. *Nodes-Sensitive Layer* ANNs: each node is specific to the position it occupies within the ANN; e.g. the nodes closest together communicate more intensely than they do with those farther away.

The Signal Flow
Every ANN signal may proceed in a direct fashion from input to output or in a more complex fashion. Thus we have two types of flow strategies:

1. *Feed forward ANN*: the signal proceeds from the input to the output of the ANN passing through all nodes only once.
2. *ANN with Feedback*: the signal proceeds with specific feedback determined beforehand or depending on the presence of particular conditions.

The ANNs with Feedback are also known as *Recurrent ANNs* and are the most plausible from a biological point of view. They are often used to process timing signals and they are the most complex to deal with mathematically. In an industrial context, therefore, they are often used with feedback conditions determined a priori in order to ensure stability.

The Learning Strategy
Every ANN can learn, over some period of time, the properties of the environment in which it is immersed or learn the characteristics of the data which is presented to the network. This is accomplished in basically one of two ways or mixture of both:

1. By reconstructing approximately the probability density function of the data received from the environment, compared with preset constraints.
2. By reconstructing approximately the parameters that solve the equation relating the input data to the output data, compared with preset constraints. The first method is known in the context of ANNs as *Vector Quantization*; the second method is known as *Gradient Descent*. The Vector Quantization method can be thought as relating the input variables to the output variables via *hyper-spheres* of a defined range. The Gradient Descent method can be thought as relating the input to the output variables via *hyper-planes*.

The difference between these two methods becomes evident in the case of a feed forward ANN with at least one hidden unit layer. With Vector Quantization the hidden units encode locally the more relevant traits of the input vector. At the end of the learning process, each hidden unit will be a prototype representing one or more relevant traits of the input vector in definitive and exclusive form. With gradient descent, the hidden units encode, in a distributed manner, the most relevant characteristics of the input vector. Each hidden unit, at the end of the learning process, will tend to represent the relevant traits of the input in a "fuzzy" and non-exclusive fashion. In summary, vector quantization develops a local learning, while gradient descent develops a distributed or vectorial learning. There are considerable differences between these two approaches:

- Distributed learning is computationally more efficient than local learning. It may also be more biologically plausible (not always or in every case).
- When the function that connects input to output is nonlinear, distributed learning may "jam up" at local minima due to the use of the gradient descent technique that is a local minimization method (rather than a global optimization method).
- At the local level, the input is more sharply defined from the output when using vector a quantization than when using gradient descent.
- When interrogating an ANN trained with vector quantization, the ANN responses cannot be different from those given during learning; in the case of an ANN trained with gradient descent the responses may be different from those obtained during the learning phase.
- These features are so important that families of hybrid ANNs treating the signal in 2 steps have been designed. The first step uses the vector quantization method and the second step uses the gradient descent method.
- Local learning helps the user understand how the ANN has interpreted and solved the problem; distributed learning makes this task more complicated (though not impossible).
- Local learning is a competitive type, nodes compete to become dominant with relatively large values while the other nodes lose and tend to zero value; distributed learning presents aspects of both competitive and cooperative behavior between the nodes where the winning node or neuron determines the spatial topological neighborhood of winning nodes thus yielding cooperation between these neighboring neurons.

2.2 Artificial Neural Network Typology

We discuss in this section, the two fundamental types of neural networks that are, in fact, one type. We are leaving a general discuss of the full range of neural network types to the last part of the chapter. Since our focus is on a particularly useful type of artificial neural network, Auto-CM, which is an unsupervised neural network and its supervised version, K-CM, supervised and unsupervised ANNs are discussed next.

Two Fundamental ANNs
Traditionally ANNs are divided into two families: *Supervised ANNs* and *Unsupervised ANNs*. Supervised ANNs utilize a known target toward which the training evolves. The weights are corrected according to the error between the output and the target. Unsupervised ANNs are used when the output is not necessarily known for a particular input. But from a theoretical point of view this distinction could be superficial. An interesting viewpoint on this theoretical debate can be gained by noting that, from the point of view of the energy function that is being calculated by an

unsupervised versus a supervised ANN, it is easy to subsume both approaches into a common framework for the case of the linear perceptron. The energy function for a supervised ANN can be written as its Mean Square Error:

$$MSE = \frac{1}{2} \sum_{p}^{K} \sum_{i}^{N} \left(t_{p,i} - u_{p,i}\right)^{2} \tag{2.1}$$

where $t_{p,i}$ and $u_{p,i}$ are the ith target and output pth pattern respectively and K and N are the number of patterns and output respectively.

Whereas, traditionally, the energy minimization function in an unsupervised auto-associative neural network is represented by the following equation:

$$En = \frac{1}{2} \sum_{p}^{K} \sum_{i}^{N} \sum_{j}^{N} u_{p,i} u_{p,j} w_{i,j} \tag{2.2}$$

where $w_{i,j}$ is the trained weights from input j to output i. Assuming that Eq. (2.1) represents the mean error of a linear perceptron, we can develop Eq. (2.1) as follows:
Let $u_{p,i} = \sum_{j}^{N} u_{p,j} w_{i,j}$

$$MSE = \frac{1}{2} \sum_{p}^{K} \sum_{i}^{N} \left(t_{p,i} - u_{p,i}\right)^{2} = \frac{1}{2} \sum_{p}^{K} \sum_{i}^{N} \left(t_{p,i} - \sum_{j}^{N} u_{p,j} w_{i,j}\right)^{2}$$

$$= \frac{1}{2} \sum_{p}^{K} \sum_{i}^{N} \left(t_{p,i}^{2} - 2t_{p,i} \sum_{j}^{N} u_{p,j} w_{i,j} + \left(\sum_{j}^{N} u_{p,j} w_{i,j}\right)^{2}\right) \tag{2.3}$$

It is clear that we can simplify (2.3) so that $MSE = \frac{1}{2} \sum_{P}^{K} \sum_{I}^{N} \left(u_{p,i}\right)^{2}$ by setting all targets to 0, as in the case of unsupervised neural networks. Thus, we have:

$$MSE = \frac{1}{2} \sum_{p}^{K} \sum_{i}^{N} \left(\sum_{j}^{N} u_{p,j} w_{i,j}\right)^{2} = \frac{1}{2} \sum_{p}^{K} \sum_{i}^{N} \left(\sum_{j}^{N} u_{p,j} w_{i,j}\right) \left(\sum_{j}^{N} u_{p,j} w_{i,j}\right)$$

$$= \frac{1}{2} \sum_{p}^{K} \sum_{i}^{N} u_{p,i} \left(\sum_{j}^{N} u_{p,j} w_{i,j}\right) \tag{2.4}$$

It is easily seen that:

$$MSE = \frac{1}{2} \sum_{p}^{K} \sum_{i}^{N} \sum_{j}^{N} u_{p,i} u_{p,j} w_{i,j} \tag{2.5}$$

and Eq. (2.5) is the energy function for an unsupervised ANN (see Eq. 2.2).

Therefore, when the target = 0,

$$En = MSE \tag{2.6}$$

Thus we can, in principle for the linear perceptron, regard unsupervised ANN learning, conceptually, as a more economical approach than supervised learning in that it entails doing away with some free parameters, namely, targets. Or, on the other hand, we can make a case for supervised learning, that is, for the inclusion of the extra free parameters, as a way to focus the learning model upon a more clear-cut task. Adopting this point of view ANNs can be classified into three sub-families:

1. Supervised ANNs;
2. Unsupervised Auto-Associative and Hetero-Associative Memories;
3. Unsupervised Autopoietic ANNs

Furthermore, the following pseudo code is a general framework to build any kind of ANN (Supervised and Unsupervised):

1) Design of the Architecture of the Network
2) Initialization of Weights
3) Do Epochs
        {
                do Cycles
                {
        a. Presentation of one Pattern as Input Vector
        b. Signal Transfer up to the Output layer
        c. Error Computation for each Node and Weight
        d. Weights and/or Nodes updating
                        Possible Recurrence
                } Until (all Patterns are presented)
        } Until (Cost Function is Optimized)

The best-known neural network is discussed next. A discussion of the various types of neural networks is given in the ensuing chapters. The following figure introduces the typology used in the ensuing chapters.

This typology as depicted in Fig. 2.2 provides a theoretical framework to understand the next chapters and paragraphs, but also it is a road map in the field of artificial neural networks. In Fig. 2.2 backpropagation (BP), support vector machines (SVM), self-organizing maps (SOM) and others ANNs are depicted.

2.2.1 Supervised ANNs

The most fundamental type of problem that an ANN must deal with is expressed as follows: *Given N independent variables about which it is easy to gather data, and M*

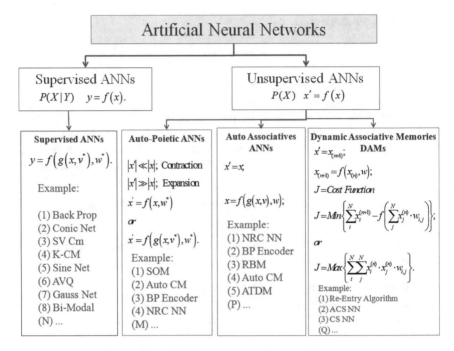

Fig. 2.2 Types of ANNs and their main characteristics

dependent variables, about which it is difficult and costly to gather data because they are the result of some operation or function performed on the N variable data, assess whether it is possible to predict the values of the M variables on the basis of the N variables. This family of ANNs is called *supervised* ANNs and their prototypical equation is:

$$y = f\left(x, w^*\right) \tag{2.7}$$

or in probabilistic terms

$$P(X|Y) \tag{2.8}$$

where y is the vector of the M variables to predict and/or to recognize (i.e. the target), x is the vector of N variables working as networks inputs, w^* are the optimized weights in a matrix of the set of parameters to approximate and f ($\cdot$) is a nonlinear function that is our resultant model. When the M variables occur in time subsequent to the N variables, the problem is described as a prediction problem. When the M variables depend on some sort of typology, the problem is described as one of *recognition* and/or *classification* and is also referred to as the proscription problem.

Conceptually, it is the same kind of problem of *using values for some known variables to predict the values of other unknown variables* via the construction of the

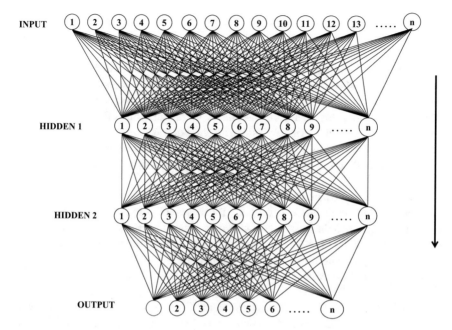

Fig. 2.3 Example of multi-layer perceptron

function. To correctly apply an ANN to this type of problem it is wise (necessary) to run a validation protocol. We start with a good sample of cases each of which has the N variables (known) and the M variables (to be discovered) that are known and reliable. This sample of complete data is needed in order to:

- *Train* the ANN;
- *Assess* its predictive performance.

The validation protocol uses part of the input sample to train the ANN (Training Set), while the remaining cases are used to assess the predictive capability of the ANN (testing set or validation set), to tune the ANN and to reveal if over fitting has occurred. In this way, we are able to test the reliability of the ANN in tackling the problem *before* putting it into operation. Now we provide some example of Supervised ANNs.

An Example of a Multi-layer Feed Forward Network: Back Propagation
Back Propagation (BP) refers to a broad family of Multi-Layer Perceptrons, whose architecture consists of different interconnected layers, see for example [1–4]. The BP ANNs represents a kind of supervised ANN, whose learning algorithm is based on the *steepest-descent* technique. If provided with an appropriate number of hidden units, they will also be able to minimize the error of nonlinear functions of high complexity (see Fig. 2.3).

Theoretically, a BP provided with a single layer of hidden units is sufficient to map any function $y = f(x)$ given that there are not many restrictive conditions. Practically, it is often necessary to provide these ANNs with at least 2 layers of hidden units when the function to compute is particularly complex or when the data chosen to train the BP are not particularly reliable.

A BP network employs a learning function that tends to "distribute itself" over the connections. This means that, in the case of BP, with at least one layer of hidden units, these units tend to *distribute among themselves* the codification of each feature of the input vector. This makes the learning more compact and efficient (number of patterns able to code in distributed hidden layers) but it is more complex to know the "reasoning" which brings a BP, in the testing process, to answer in a certain way. In brief, it is difficult to explain the *implicit knowledge* that these ANNs acquire in the training process.

A second theoretical and operative difficulty that BP poses concerns the *minimum number* of hidden units that are necessary for these ANNs to compute a function. In fact, it is known that if the function isn't linear, at least one layer of hidden units will be necessary. But, at the moment, stating exactly the minimum number of hidden units needed to compute a nonlinear function is unknown. Therefore, for this monograph, the number of hidden units is based on experience and on some heuristics.

Experience advises us to use a minimum number of hidden units the first time training of an ANN. If the training succeeds, an analysis of the sensitivity will normally allow us to understand the number of input nodes that determine the given output. Consequently, it will be able to deduce the degrees of freedom needed by the ANN to resolve the equation, that is, the construction of the unknown function f that relates the input to the output and then to express the latter in the form of the number of hidden units.

This procedure isn't guaranteed to succeed. During the training process, the BP can become trapped in a local minimum. This is because of the relationship between the morphological complexity of the hyper surface that characterizes the function that the ANN seeks to construct and the weight values, which are randomly set before the training.

The dilemma of BP is that for a prior, unknown minimum number of hidden units useful to compute a function, if *too many* units are inserted, the BP will over-fit the function and the results are likely not useful. On the other hand, if *not enough hidden layers* are created, the BP will likely have difficulty learning either because the function is too complex or because the BP randomly falls into a local minimum. The BP family includes both Feed Forward ANNs and Feedback ANNs (also known as Recurrent Networks, see for example [3]).

Let's next define the *Forward Algorithm*. The basic rule in order to calculate the activation value of a unit with respect to other units which are connected to it, having a strength w_{ji}, is a function of the weighted sum of the Inputs:

$$u_j = f\left(\sum_i u_i w_{j,i}\right) = f\left(Net_j\right)$$

$$= \frac{1}{1 + e^{-Net_j}} \tag{2.9}$$

where Net_j = Net Input to the j level unit, that is, $Net_j = \sum_i u_i \cdot w_{ji}$.

One must add to this equation the *threshold* of the unit, described as the *inclination* of the unit to activate or inhibit itself. This means that:

$$u_j = f\left(Net_j\right) = \frac{1}{1 + e^{-\left(\sum_i w_{j,i} u_i + \theta_j\right)}} \qquad (2.10)$$

where, $Net_j = \sum_i u_i \cdot w_{ji} + \theta_j$ and θ_j, is the *bias* of unit u_j; this is the degree of *sensitivity*, with which the unit u_j answers to the perturbation it receives by the net input. The bias is the opposite of the threshold and it behaves as a weight generated by a fixed input of unit value. Equation (2.9) represents the forward algorithm of BPs, assuming as default, the sigmoidal function. It is necessary to calculate the dynamic of the back propagation or correction algorithm. The mathematical basis of this algorithm was already substantiated in [4, 5] through what is called the *Delta Rule*, a procedure that allowed to correct for excess or defect of the weights between the network units, on the basis of the difference between the actual output and the desired target. Nevertheless, the delta rule allows correction of only those weights that connect the output units with those of the just underlying layer. It doesn't allow one to know in a 3-layer network, how at each cycle, the weights that connect the input units with the hidden units, should be modified. Let's examine in detail the delta rule. It is worth noting that in the ANN literature the input is usually not counted as a layer.

The coefficient of error in this procedure is calculated by considering the difference between the actual output and the desired one (the target one) and relating this difference to the derivative between the activation state of the actual output and the net input of that output.

From Eq. (2.9) we have

$$\frac{\partial u_j}{\partial Net_j} = u_j\left(1 - u_j\right). \qquad (2.11)$$

Then the Δout_j error coefficient will be,

$$\Delta out_j = \left(t_j - u_j\right) u_j \left(1 - u_j\right) \qquad (2.12)$$

where: t_j = desired Output (Target); u_j = actual Output; $u_j\left(1 - u_j\right)$ = derivative between actual Output and Net Input of unit u_j. This is based on the fact that:

$$\Delta w_{j,i} = -\frac{\partial E_p}{\partial w_{j,i}} \qquad (2.13)$$

$$E_p = \frac{1}{2} \sum_k \left(t_{p,k} - u_{p,k}\right)^2 = \frac{1}{2} \sum_k \left(t_{p,k} - f_k\left(Net_{pk}\right)\right)^2$$

$$= \frac{1}{2} \sum_k \left(t_{p,k} - f_k\left(\sum_j w_{k,j} u_{p,j} + \theta_k\right)\right)^2 \tag{2.14}$$

where: E = error; p = pattern; t_k = target; u_k = output. And then we have,

$$\frac{\partial \left(Net_{pk}\right)}{\partial w_{k,j}} = \frac{\partial \left(\sum_j w_{k,j} u_{p,j} + \theta_k\right)}{\partial w_{k,j}} = u_{p,j} \tag{2.14a}$$

$$-\frac{\partial \left(E_p\right)}{\partial w_{k,j}} = \left(t_{p,k} - u_{p,k}\right) f'_k\left(Net_{pk}\right) u_{p,j} \tag{2.14b1}$$

where

$$f'_k\left(Net_{pk}\right) = \frac{\partial f_k}{\partial Net_{p,k}} = \frac{\partial u_{p,k}}{\partial Net_{p,k}} = u_{p,k}\left(1 - u_{p,k}\right) \tag{2.14b2}$$

One can say that the quantity to be added or subtracted from weight w_{ji} will be decided by value Δout_j, with respect to the activation state of unit u_j namely, the activation with which u_j is connected to weight w_{ji} and in relation to the coefficient r, the learning rate. This is the correction rate that one wants to adopt (when $r = 1$, then the value of the adding or subtracting from weight w_{ji} is the one calculated by the whole procedure). The delta rule is depicted in Eqs. (2.14a, 2.14b1, 2.14b2 and 2.15) above and below.

$$\Delta w_{j,i} = r \cdot \Delta out_j \cdot u_j \tag{2.15}$$

The value Δw_{ji} can be both negative and positive. It represents the "quantum" to be added or subtracted from the previous value of weight w_{ji}. Then,

$$w_{j,i(n+1)} = w_{j,i(n)} + \Delta w_{j,i} \tag{2.16}$$

Nevertheless, Eq. (2.12) presupposes that each arriving unit of a weight has an actual value, which is comparable with an ideal value, towards which it should tend (Target). This presupposition, however, is valid only for the weights that connect a unit layer with the layer of the output units.

The correction procedure discussed till now, related only to BP provided with 2 layers (recall the input is not usually considered a layer). The delta rule, then, represented by Eq. (2.15), allows one to carry out the weight's correction only for very limited networks. For *multilayers* BP, this is, with 1 or more hidden layers, the delta rule, is insufficient. For example,

Fig. 2.4 Delta rule

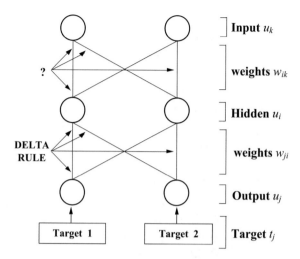

In Fig. 2.4 it is evident that, while the correction of weights w_{ji} is possible through the Delta Rule, because the value that the units u_j should assume (t_j) is known, the correction of weights w_{ik} is not possible. This is so because an ideal reference does not exist for the unit u_i. In fact, the value that they are going to assume is one of the *results* of BP's learning and therefore it can't be constrained. Rumelhart and others solve this problem through a *generalization* of the traditional delta rule (see [6]).

The generalization of the delta rule consists in modifying Eq. (2.12), in those cases in which the weight to modify isn't connected to an output unit. Therefore, instead of computing the difference between the actual output and the desired one, a report summation will be computed between the error coefficient Δout_j, previously calculated, and the weights which that coefficient was referring to:

$$\Delta hidden_i = u_i \cdot (1 - u_i)$$
$$\cdot \sum_j \left(\Delta out_j \cdot w_{j,i} \right) \tag{2.17}$$

and therefore,

$$\Delta w_{i,k} = r \cdot \Delta hidden_i \cdot u_k \tag{2.18}$$
$$w_{i,k(n+1)} = w_{i,k(n)} + \Delta w_{i,k} \tag{2.19}$$

In fact, starting from Eq. (2.14) we have that:

$$\frac{\partial (E_p)}{\partial w_{k,j}} = \frac{1}{2} \sum_k \frac{\partial}{\partial w_{k,j}} \left(t_{p,k} - u_{p,k} \right)$$
$$= -\sum_j \left(t_{p,k} - u_{p,k} \right) f_k \left(Net_{pk} \right) w_{i,j} f_j \left(Net_{pi} \right) u_{p,i} \tag{2.20}$$

where w_{kj} = hidden-outputs weights. Through the generalized delta rule it is possible to create a back propagation algorithm, capable of correcting the weights of any BP's layer at every cycle. We can now synthesize the two algorithms through which the BP would be able to work:

a. **Forward Algorithm**

Forward algorithm is given by (see Eq. 2.10):

(1) $Net_j = \sum_i w_{j,i} \cdot u_i + \theta_j$
(2) $u_j = f\left(Net_j\right)$

b. **Back Propagation Algorithm**

b1. *Correction calculation of the weights connected to the Output* (see Eqs. (2.12) and (2.15)):
(1) $\Delta out_i = \left(t_j - u_j\right) f'\left(u_j\right)$
(2) $\Delta w_{j,i} = r \cdot \Delta out_j \cdot u_j$

b2. *Correction calculation of the weights not connected to the Output* (see Eqs. (2.17) and (2.18)):
(1) $\Delta hidden_i = f'(u_i) \cdot \sum_j \Delta out_j w_{j,i}$
(2) $\Delta w_{i,k} = r \cdot \Delta hidden_i \cdot u_k$

b3. *Fulfillment of the corrections on the weights* (see Eqs. (2.16) and (2.19), and see Fig. 2.4 for weight ($w_{j,i(n)}$ and $w_{i,k(n)}$) and change in weight ($\Delta w_{j,i}$ and $\Delta w_{i,k}$)):
(1) $w_{j,i(n+1)} = w_{j,i(n)} + \Delta w_{j,i}$
(2) $w_{i,k(n+1)} = w_{i,k(n)} + \Delta w_{i,k}$

At this point, both the *minimum conditions of activation* and those of *functioning*, or learning for BPs, have been explained.

2.2.2 Unsupervised ANNs

Unsupervised ANNs are algorithms that approximate from data, a function in a manner that is presented next.

$$x' = f(x, w) \qquad\qquad (2.21)$$

or in probabilistic term $P(x)$

where:

x the input vector
w the weights to approximate the function or another function $g(\cdot)$
x' a projection of the input vector in another space.

The term "autopoietic" is used in relation to ANNs to refer to neural networks that are reproducing and self-maintaining.

(a) **Autopoietic** $\left(|x'| \neq |x|\right)$—In this case the output of the ANN is a projection of the N dimensional input space onto the P dimensional space of the ANN output. When $P \ll N$ we have the classic case of multi-dimensional scaling or the nonlinear principal components extraction (it is the same work that the function $g(\cdot)$ does in Eq. (2.21b2)). Self-organizing maps (SOMs) (see [7]) and complex auto-associative ANNs (see [6, 8]), with their hidden layer, can represent a case of nonlinear component analysis. When $P \gg N$, that is the projection output space is bigger that the input space, we have multi-dimensional expansion. This is partially what auto-associative ANNs use when with their weights w (see Eq. (2.21b1)) to decode the hidden units into a bigger output space. But the only ANN able to project *directly* the input space (N) onto the bigger space of its weights ($N \times N$) is Auto-CM, the ANN to which this book is dedicated. Then we can rewrite Eq. (2.21) in the following way:

$$x' = f(x, w) \tag{2.21a}$$

where

if $|x'| < |x|$ space compression
if $|x'| > |x|$ space expansion

(b) **Auto-Associative** $\left(x' = x\right)$—In this case the output of the ANN tries to learn its own input. These kinds of ANNs tend to approximate the implicit function of the data. In other words: *given N variables defining a dataset, find its optimal connections matrix that is able to relate each variable in terms of the others and consequently to approximate the hyper-surface on which each data-point is located. That is, find the functional relationship, its hyper-surface, which is imputed by the dataset.* We can rewrite Eq. (2.21) in two different forms.

$$x = f(x, w), \quad w_{i,j} = 0. \tag{2.21b1}$$

In this case we have a *simple auto-associative (SAA)* ANN with no hidden units. The second form is:

$$x = f(g(x, v), w), \tag{2.21b2}$$

where:

$g(x, v)$ function for hidden units with its encoder weight (v);
$f(g(x, v), w)$ output function with its decoder weight (w).

In this second case (2.21b2) we have a complex auto-associative (CAA) ANNs, also referred to as an encoder. Encoders are often used to initialize the weights matrices of the deep supervised ANNs (see [8]). A simple auto associative ANN can be represented by a simple Auto Associative BP without hidden units. A complex Auto Associative ANN can be represented by an Auto-Associative BP with at least one layer of hidden units [9] or by a New Recirculation ANN [7] or also by Restricted Bolzmann Machine [10]. These CAA ANNs also can work as AutoPoietic ANNs, because they project the input space in their hidden units. Auto-Associative has been used successfully in face recognition, signal processing and medical imaging [11–13].

(c) **Hetero-Associative** $(y' = f(g(x, y, v), w))$—In this case the output of the ANN tries to recall an output that is different but related to the input. In other words, the ANN output associates data from a category that is related to data of another category:

$$y = f(x, w), \quad w_{i,j} = 0. \tag{2.21c1}$$

In this case we have a *simple hetero-associative memory* ANN with no hidden units. A second form is:

$$y = f(g(x, v), w), \tag{2.21c2}$$

where $g(x, v)$ is the function for the hidden units with its encoder weight v and $f(g(x, v), w)$ is the output function with its decoder weight w.

(d) **Dynamic Associative Memories (DAM)** (see for example [11, 12]), $(x^{[n+1]} = h(x^{[n]}))$. In this case Eq. (2.21) can be written in the following way:

$$x^{[n+1]} = f\left(x^{[n]}, w\right) \tag{2.21d}$$

where

$x^{[n=0]}$	the original input vector;
$x^{[n]}$	the input vector at the nth cycle;
$x^{[n+1]}$	the input vector at the $(n + 1)$th cycle;
w	the weights to approximate the function vector.

We need to distinguish two different types of Dynamic Associative Memories.

(i) **DAM with Energy Minimization**

These DAMs minimize the energy of the output according to the constraints of the weights and of the external input. In this case the ANN algorithm is simply a way to navigate the hyper surface approximated by the weights matrices during the training phase. These ANNs use fuzzy outputs to define the optimal value of each variable when a specific combination of external input is activated as a set of constraints. DAMs that follow the energy minimization criterion are particularly suitable to simulate dynamic scenarios such as after that a Complex Auto-Associative ANN that was trained on a data set. Which values all the variables tend to take when some of these variables are set up with specific fixed values? And which records are proportionally activated in this context? An example of this kind of ANN is a Complex Auto Associative ANN whose weights are explored using the Re-Entry algorithm (see [10]).

(ii) DAM with Energy Maximization

These ANNs try to maximize the output using weights and the external input as simple constraints. They use excitatory and inhibitory weights and an external not null input as constraints, in order to maximize each variable activation state. They are a good example of Content Addressable Memory (see [1, 2, 4–6]).

We next present one representative ANN of each of these three types.

An Example of Auto-Poietic ANN: Self Organizing-Map

A Self-Organizing Map (SOM) is an example of an autopoietic neural network, which is our next topic. The SOM is a neural network attributed to Kohonen [7, 14] who developed it between 1979 and 1982. It is an unsupervised type of network which offers a classification of the input vectors creating a prototype of the classes and a projection of the prototypes on a two-dimensional map (but n-dimensional maps are also possible) able to record the relative proximity (or neighborhood) between the classes. This means that the network offers important synthetic information about the input, which we describe next. A SOM does the following things:

a. Classifies the input vectors on the basis of their vector similarity and assigns them to a class;
b. Creates a prototypical model of the classes with the same cardinality (number of variables) as the input vector;
c. Provides a measurement, expressed as a numerical value, of the distance/proximity of the various classes;
d. Creates a relational map of the various classes, placing each class on the map itself;
e. Provides a measurement of the distance/proximity existing between the input vectors and the class to which they have been assigned and between the input vectors and the other classes.

The relative simplicity of the network architecture led to its successful application and allowed its implementation to be replicated and verified. A typical SOM network is made up of 2 layers of units, a one-dimensional input (n-dimensional vector) and a two-dimensional output layer (rows (r) × columns (c)), also known as Kohonen's map (matrix M of dimension r × output layer and each unit of the input layer as a tensor matrix W of dimension $r \times c \times n$. The weight vector connecting each output unit to an input unit is called a "codebook" (a "vector" of W of cardinality n, see Fig. 2.5). Within the SOM network each output unit can be interpreted as a class whose codebook represents the prototype.

The SOM algorithm (Fig. 2.6) is based on a competitive algorithm founded on the vector quantification principle as follows. At each cycle (time, iteration) in the network, the unit from Kohonen's layer whose codebook is most similar to the input wins. This unit is named "winner unit" (WU). Consequently, the WU codebook is

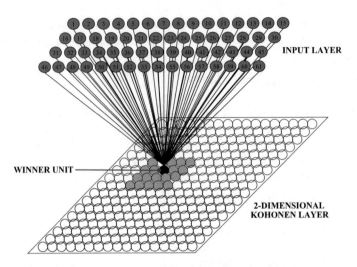

Fig. 2.5 Example of unsupervised ANN for natural clustering—self-organizing map

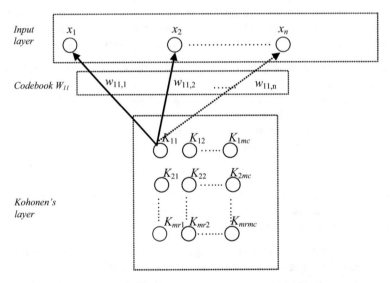

Fig. 2.6 SOM with n-nodes of input, with $(m_r \Delta m_c)$ units of Kohonen's layer. This architecture allows the inputs to be classified into m^2 classes, each being a subclass represented by a codebook

modified to get it even closer to the input. The codebook belonging to the units that are physically near the WU (which are part of the neighborhood) are also put closer to the input by a given delta adjustment.

The algorithm calculates a first stage during which the parameters of a neighborhood and corrections of weights are set and the codebook initialization is carried out; this stage is followed by the cyclic stage of codebook adjustment. At this stage

the codebook is modified for the network to classify the input records. In short, the SOM algorithm is organized as follows:
   Initialization stage

- Layering of the input vectors;
- Definition of the dimensions (rows × columns) of the matrix which, in its turn, determines the number of classes and therefore of prototypes (codebook);
- Initialization of the codebook whose values are randomly chosen;
- Definition of the function (Gaussian, Mexican hat, etc.) and of the parameters regulating the neighborhood of the WU and of the weight correction, the delta.

   Cyclic calibration stage

- Presentation of the input vectors (pattern) in a random and cyclic way to the network;
- Calculation of the d-activation of the K units of Kohonen's layer. The activation is calculated as the vector distance between the input vector X and the weight vector W_j (from the $m j$th codebook) which links the K units to the input nodes. The classic way is to calculate the Euclidean distance between the vectors,

$$d_j = \|X - W_j\| = \sqrt{\sum_{i=1}^{N} (x_i - w_{i,j})^2} \tag{2.22}$$

- Calculation of the winner unit, WU. The node of the Kth layer whose activation is less than

$$WU: d_W = \min_{j \in [1,M]} \left\{ d_j = \|X - W_j\| = \sqrt{\sum_{i=1}^{N} (x_i - w_{i,j})^2} \right\} \tag{2.23}$$

is the winner.
- Correction of the codebook (matrix of the W_{ij} weights) of the winning unit and the units adjacent to the winning unit in relation to the function set to determine the level of weight correction, according to the input and the proximity to the WU;
- Updating of the factors determining the proximity and layering of the delta correction of the codebooks.

   The distinctive characteristic of the SOM is mainly related to the updating of the weights, carried out not only on those related to the WU but also, according to the chosen function, on the weights belonging to the units which are physically close to it. This characteristic also allows the SOM to show the position occupied by the class within the matrix in relation to the position occupied by the other classes. This type of topological mapping, able to organize the classes through spatial relations, has been given the name of *Feature Mapping*.

   The Topology of the Neighborhood: The neighborhood of a WU is defined by the value of physical proximity (v) existing between the WU and the other K units.

Each unit of Kohonen's layer occupies a position on the matrix of the coordinates (r, c), for which the neighborhood is indexed with a scalar degree from 1 to the maximum row and column dimension,

$$v_i = \pm r \quad OR \quad v_i = \pm c \quad where \quad \max i = \max r \quad OR \quad \max c. \tag{2.24}$$

A function $h(v)$ is defined and it regulates the size of the neighborhood and the extent of the corrections which need to be made to the codebook of the units close to the WU. With the passing of time (the cycles) the neighborhood is reduced until it disappears. In this case the only codebook of the winner unit will remain to be updated. Since the codebook is set during the initialization stage with random values within the layering range, the proximity/distance of the WU at the beginning of the learning stage is regulated by the maximum size in order to allow for all the elements of the codebook to be modified and put closer to the input vectors. The reduced distance, whose matrices have more columns than rows (are wide), occur because some areas of the K matrix remain isolated since the codebook is significantly different from the input vectors. Function $h(v)$ must also allow for the correction, to increase for the units close to the WU, and therefore to decrease when v is larger. The Gaussian function,

$$h(v) = e^{\frac{-v^2}{\sigma}} \tag{2.25}$$

has been shown to meet these needs remarkably well, where v is the physical proximity, that is, distance, of the unit to the winner unit, WU, and σ is a parameter that linearly decreases by a Δ as time increases, thereby modifying the width of the bell curve. This affects the shape (extent) of the classification neighborhood. Figures 2.7 and 2.8 show examples of neighborhood space topologies.

Correction of the codebook
The rate at which correction a codebook undergoes is determined by various factors. These are:

a. The difference (d) existing between the vector codebook and the input vector;
b. The physical distance to the WU (v);
c. The function value of the neighborhood $h(v)$ that determines a $\Delta\sigma$;
d. The function value of weight layering in relation to the iteration count of the network that determines a $\Delta\alpha$.

The SOM *codebook* is moved closer to the input vector where "closer" means smaller in norm. Therefore, for each generic *codebook*, W, the distance existing between the corresponding weights w_{ij} and the variables x_i of the generic input vector X is calculated by,

$$d_j = \|X - W_j\| = \sqrt{\sum_{i=1}^{N} (x_i - w_{i,j})^2} \tag{2.26}$$

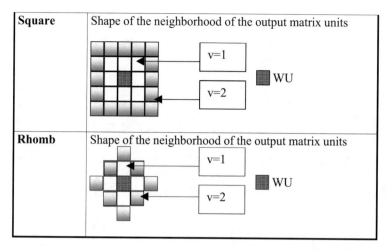

Fig. 2.7 Topology of the neighborhood Space of a *WU* in a square and in a rhombus. In the illustration *v* is the degree of proximity of the K units to the *WU*

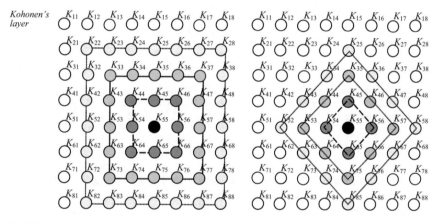

Fig. 2.8 An example of the topology of the neighborhood space with matrix K ($8r \times 8c$) where the *WU* is the K_{55} unit. The first matrix shows a neighborhood in a square while the second a neighborhood in a rhombus. We can notice from the illustration that, for example, while in the matrix to the left the v distance of the K_{66} unit to the *WU* is 1, in the matrix to the right the v distance of the K_{66} unit to the *WU* is 2

On the basis of the function $h(v)$ of the neighborhood, the $\Delta\sigma$ is calculated in relation to the value of the parameter σ and the proximity (v) of the unit K to the *WU*. $\Delta\sigma$ is the measure which is assumed in the function $h(v)$ when $x = v$. In the case in which function $h(v)$ is the *Gaussian curve*, the $\Delta\sigma$ will be calculated in the following way (see Fig. 2.9)

$$\Delta\sigma = h(v) = e^{\frac{-v^2}{\sigma}} \tag{2.27}$$

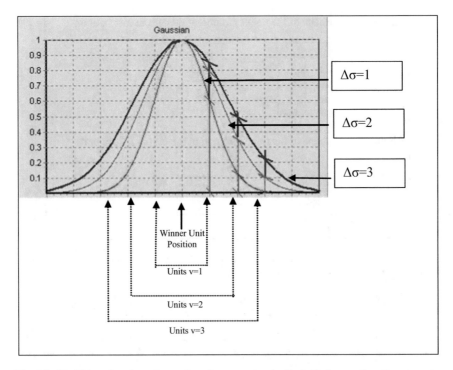

Fig. 2.9 The illustration shows how, when the parameter $\Delta\sigma(1, 2, 3)$ changes then the parameter that determines the correction curve of the neighborhood function, the number of units that are part of the neighborhood and the extent of the correction ($v1$, $v2$, $v3$) made on the weights, also change $\Delta\sigma = e^{\frac{-v^2}{\sigma}}$ for $\sigma = 1, 2, 3$

The $\Delta\alpha$ is calculated as a factor of a linear function, decreasing in relation to the total number of iterations associated with the network. Therefore, the function of correction of the *codebook* is,

$$f(w) = \alpha e^{\frac{-v^2}{\sigma}} \sqrt{\sum_{i=1}^{N} (x_i - w_{i,j})^2} \tag{2.28}$$

and

$$w_{i,j}^{(m+1)} = w_{i,j}^{(m)} + \alpha(m) e^{-\frac{v^2}{\sigma(m)}} (x_i - w_{i,j}). \tag{2.29}$$

2.3   A Few Final Theoretical Considerations

It is now useful, in the context of ANNs, to discuss the following:

a. The place that ANNs have in the general framework of the natural sciences;
b. What are the basic (atomic) features needed to construct a neural network.

ANNs belong to the field of Artificial Adaptive Systems (AAS). AAS is a new and a special branch of Natural Sciences. AAS are the new computerized laboratories by means of which researchers try to simulate natural processes in order to explain their complex and hidden laws. More specifically, ANNs are a special type of computerized laboratories able to reproduce adaptive natural processes. From this perspective, data can be thought to represent a statistical sample of the natural process we intend to understand.

ANNs have had three different objectives, from an historical point of view:

a. To understand how the human brain works by means of its artificial simulation. The comprehension of human brain physiology is the main goal of this objective and ANNs are important **simulation** tools in these approaches;
b. To develop new computation algorithms, inspired to human brain architecture, able to processes information, more effectively and expeditiously than otherwise possible. The main goal of this aspect of ANNs is to define new algorithms, and the human brain structure represents a model to emulate.
c. To understand in every natural process how the transformation from individual behaviors to collective behaviors work. In other words, how the transition from the simple and local processes to the global and complex processes work in nature. In this case the main target of the scientist is the discovery of natural laws and the ANNs represent a set of new algorithms with which we can implement the correct experiments to verify our hypothesis. In this case ANN algorithms represent a powerful experimental framework for science.

These three different aspects, especially the last one, need a general theory able to explain how to build these algorithms and/or tools in a mathematical and correct way that mimics the actual system as closely as possible. For these reasons we think that ANNs are useful techniques that need to be incorporated into a general theory of artificial adaptive systems.

It is therefore important to elaborate a framework able to assist us in the construction of new ANNs, starting from an assembly of atomic components. For this reason we propose a bottom-up theoretical process composed of three components (a basic layer, a central layer, and a complex layer) and two sub-components for each step, its semantics/structure and its syntax/function.

The basic framework that was presented in this chapter we give below in outline form.

1. Basic layer

    i.   Structure (syntax): Nodes
    ii. Function (semantics): Connections
    iii. Result: Networks

2. Central Layer

    i.   Structure (syntax): Networks
    ii. Function (semantics): Processors that update/change the connections according to various laws of learning, local signal flow and other constraints
    iii. Result: Learning and Evolving Networks

3. Complex Layer

    i.   Structure (syntax): Learning and Evolving Networks
    ii. Function (semantics): Global signal flow rules among networks and their interaction under specific constraints
    iii. Result: Artificial Organism or Meta Algorithm

For this exposition, this framework suffices for now.

Specific References

1. Buscema, Massimo, and Pier Luigi Sacco. 2013. GUACAMOLE: A New Paradigm for Unsupervised Larning. In *Data Mining and Applications Using Artificial Adaptive Systems*, ed. Bill Tastle (pp 211–230). New York: Springer Science+Business Media.
2. Hornik, Kurt. 1989. Multilayer Feed Forward Networks are Universal Approximators. *Neural Networks* 2: 359–366.
3. Pineda, F.J. 1988. Generalization of Backpropagation to Recurrent and Higher Order Neural Networks. In *Neural Information Processing Systems*, ed. D.Z. Anderson, 602–611. New York: American Institute of Physics.
4. Werbos, P. 1974. *Beyond Regression—New Tools for Prediction and Analysis in the Behavioral Sciences*. PhD thesis, Harvard University.
5. Rumelhart, D., G. Hinton, and R. Williams. 1986. Learning Internal Representations by Error Propagation. In Rumelhart, McClelland, 318–362.
6. Rumelhart, D.E., and J.L. McClelland (eds.). 1995. *Parallel Distributed Processing*, vol. 1, 318–362. Boston: The MIT Press; Y. Chauvin, and Rumelhart, D.E. 1995. *Back Propagation. Theory, Architecture and Applications* (Chapter 1). Hillsdale, New Jersey, USA: Lawrence Erlbaum Associates.
7. Kohonen, Teuvo. 1982. Self-Organized Formation of Topologically Correct Feature Maps. *Biological Cybernetics* 43 (1): 59–69.
8. Bengio, Y. 2009. Learning Deep Architectures for AI. *Foundations and Trends in Machine Learning* 2 (1): 1–127.
9. Ko, Hanseok, and Robert H. Baran. 1994. Signal Detectability Enhancement with Auto-Associative Backpropagation Networks. *Neurocomputing* 6 (2): 219–236.
10. Marcus, C.M., and R.M. Westervelt. 1989. Dynamics of Iterated-Map Neural Networks. *Physical Review A* 40 (1): 501–504.
11. Cichocki, A., and R. Unbehauen. 1993. *Neural Networks for Optimization and Signal Processing*. Chichester: Wiley.

12. Newman, F.D., and H. Cline. 1993. A neural network for optimizating radiation therapy dosage. In *The International Conference on Numerical Analysis and Automatic Result Verification*, February 1993.

13. Raff, U., and F.D. Newman. 1992. Automated Lesion Detection and Quantitation Using Autoassociate Memory. *Medical Physics* 19 (1): 71.

14. Kohonen, T. 1989. *Self Organization and Associate Memory*. Berlin: Springer.

Chapter 3
Auto-contractive Maps

Abstract This chapter focuses on Auto-Contractive Maps, which is a particularly useful ANN. Moreover, the relationship between Auto-Contractive Map (Auto-CM), which is the main topic of this monograph, its relationship to other ANNs and some illustrative example applications are presented.

3.1 Introduction

Auto-CM is useful in solving problems in which one wishes to find relationships among data, especially in partitioning data into clusters of related elements [1] . This version of Auto-CM is unsupervised. Chapter 6 presents a supervised version of Auto-CM. The chapter looks at Auto-CM with illustrative examples that show the power of Auto-CM. Chapter 7 is devoted to comparisons of Auto-CM and other ANNs. We reiterate that not only is Auto-CM powerful just by itself, there are two additional aspects that makes Auto-CM even more powerful and useful, the visualization as a graph theoretical depiction of interrelationships uncovered by Auto-CM (found in Chap. 4) as well as the transformation of the database itself to elicit even deeper relationships (found in Chap. 5).

Auto-CM has many features that are different from classic ANNs:

1. Auto-CM process all the variables together finding the many-to-many relationships among them;
2. All its weights, before the training phase, are initialized with the same value (0.00000001) and not randomly. This means that Auto-CM is a complete deterministic algorithm: processing the same data set we will get always the same results;
3. Auto-CM can treat both Boolean and real number data, also mixed in the same dataset. The data only needs to be linearly scaled, at the beginning, to between 0 and 1;
4. Auto-CM can be used both as an unsupervised ANN as well as its supervised version (see K-CM, Chap. 6);
5. The supervised version of Auto-CM (K-CM) has been shown not to be affected by the phenomenon of overfitting during the training phase;

© Springer International Publishing AG, part of Springer Nature 2018
P. M. Buscema et al., *Artificial Adaptive Systems Using Auto Contractive Maps*, Studies in Systems, Decision and Control 131, https://doi.org/10.1007/978-3-319-75049-1_3

6. The Auto-CM output is irrelevant, at the end of the training phase since all the output vectors will be close to zero. It is the weights, that at beginning were all initialized with the same value close to zero, which have encoded in them the key information present into the dataset;

7. The final weights of a trained Auto-CM are organized into a non-symmetric square matrix, where each variable shows a specific value of the nonlinear correlation with any other variable, where all these associations can be expressed using fuzzy set theory. In addition they can be projected in various forms of weighted, direct or indirect graphs;

8. Auto-CM is not sensitive to the ratio between the number of variables, N, and number of records, P;

9. Auto-CM learns quickly and correctly when the number of variables, N, is much bigger then the number of records, P (N ≫ P), which means that Auto-CM can easily learn both the given data set and its transposition;

10. Auto-CM can learn the many-to-many relationships between variables and records, projecting with the same metric into a weighted graph;

11. The final matrix of weights of Auto-CM is a tensor matrix. In fact, if Auto-CM learns a set of completely orthogonal input vectors, then its final weights matrix will be null everywhere except on the main diagonal;

12. The final matrix of weights generated by Auto-CM violates all the axioms of the concept of distance, its weights are asymmetric, they do not respect the triangular inequality, and the distance between a variable from itself may bigger than zero, thereby behaving as a matrix of tensors among the database variables;

13. The Auto-CM learning process does not risk being trapped in local minima as we shall see;

14. The Auto-CM trained weights may be used to generate a robust Markov machine defining the correct probability chains of transition between all the variables;

15. Auto-CM provides a new type of energy minimization. While the usual minimization function is

$$E = Min \left\{ \sum_{i}^{N} \sum_{j}^{N} \sum_{q}^{M} x_i^q x_j^q \sigma_{i,j} \right\}; \quad \sigma_{i,j} > 0, \tag{3.1}$$

Auto-CM minimizes the energy extracted from the data set according to the following equation:

$$E = Min \left\{ \sum_{i}^{N} \sum_{j \neq i}^{N} \sum_{k \neq j \neq i}^{N} \sum_{q}^{M} x_i^q x_j^q x_k^q A_{i,j} A_{i,k} \right\}, \tag{3.2}$$

where

x_i^q = value of i-th input unit at q-th pattern;
$A_{i,j} = 1 - \frac{w_{i,j}}{C}; A_{i,k} = 1 - \frac{w_{i,k}}{C}$

$w_{i,j} < C, w_{i,k} < C$
i, j, k = indices for the variables—columns;
q = index for the patterns—rows;
N = number of variables (columns);
M = number of patterns (rows);

16. Auto-CM, in many trials, has been shown to be the best algorithm able to detect weak and strong similarities among data;
17. The final Auto-CM weights represented by its parameters can be used to construct a continuous hyper-surface of whole data set;
18. Auto-CM belongs to any of the following three typological types that we have classified as unsupervised ANNs.

    a. Auto Poietic: Auto-CM project each input vector in an expanded space of its weights;
    b. Auto Associative: Auto-CM weights define the hyper surface of the data set;
    c. Dynamic Associative Memories (DAMs): Auto-CM weights may be used for dynamic scenarios simulation;

19. Auto-CM is biologically plausible since at the beginning of the training process the input vector are repeated in the output layer (memorize the dataset) and after a while each output vector is the union of all the input vectors of the data set (detect the strong similarities); in a third logical step, the output starts to show only the outputs that each input vector presents as different from the other input vectors (detect the differences); toward the end of the training, the outputs become null (complete internalization of the data set into the weights).

3.2 The Auto-contractive Map Structure

Auto-CM is characterized by a three-layer architecture: an *input layer*, where the signal is captured from the environment, a *hidden layer*, where the signal is modulated inside the Auto-CM, and an *output layer*, through which the Auto-CM feeds back into the environment based on the stimuli previously received and processed (see Fig. 3.1).

Each layer contains an equal number of N units, so that the whole Auto-CM is made of $3N$ units. The connections between the Input and the Hidden layers are mono-dedicated, whereas the ones between the Hidden and the Output layers are fully connected. Therefore, given N units, the total number of the connections, Nc, is given by:

$$Nc = N(N + 1).$$

(3.3)

Fig. 3.1 An example of an
Auto-CM with N = 4

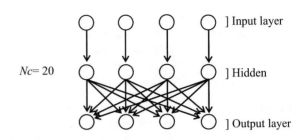

All of the connections of Auto-CM may be initialized either by assigning the same constant value to each, or by assigning values at random. The best initial value, in practice, is to initialize all the connections with a same, positive value, close to zero.

The learning algorithm of Auto-CM may be summarized in a sequence of four characteristic steps:

1. Signal transfer from the input into the hidden layer;
2. Calculation and accumulation of deltas (weights changes) for each input-hidden connection;
3. Adaptation of the values of the connections between the input and the hidden layers at the end of each epoch;
4. Signal transfer from the hidden into the output layer;
5. Calculation and accumulation of deltas for each hidden-output connection;
6. Adaptation, updating, of the value of the connections between the hidden and the output layers at the end of each epoch.

The units of the input layer are denoted $m^{[s]}$ (input layer sensors), scaled between 0 and 1, $m^{[h]}$ denotes the hidden layer units, and $m^{[t]}$ denotes the output layer units (the system targets). Moreover, we define v to be the vector of the mono-dedicated connections. W denotes the matrix of the connections between the hidden and the output layers where p will be the number of the training patterns, $p \in M$ (where M is the total number of the patterns), and n is the epoch number. A new epoch begins when all the patterns are processed. In order to specify the steps 1–6 that define the Auto-CM algorithm, we have to define the corresponding signal forward-transfer equations and the learning equations, as follows.

a. Signal transfer from the input to the hidden layer:

$$m^{[h]}_{i,p(n)} = m^{[s]}_{i,p(n)} \left(1 - \frac{v_{i(n)}}{C}\right) \tag{3.4}$$

where C is a positive real number not lower than 1, which we will refer to as the contraction parameter (see below for comments).
b. The updating of the connections, $v_{i(n)}$, is accomplished through the variation, $\Delta v_{i(n)}$, which can be thought to be like trapping the "energy" difference generated according to Eq. (3.4):

$$\Delta v_{i(n)} = \sum_{p}^{M} \left(m_{i,p(n)}^{[s]} - m_{i,p(n)}^{[h]} \right) \left(1 - \frac{v_{i(n)}}{C} \right) m_{i,p(n)}^{[s]} \tag{3.5}$$

$$v_{i(n+1)} = v_{i(n)} + \alpha \cdot \Delta v_{i(n)} \tag{3.6}$$

where α is the learning coefficient; where $\alpha \geq 1$.

c. Signal transfer from the hidden to the output layer:

$$Net_{i,p(n)} = \frac{1}{N} \sum_{j=1}^{N} \left(m_{i,p(n)}^{[h]} \right) \left(1 - \frac{w_{i,j(n)}}{C} \right); \tag{3.7}$$

$$m_{i,p(n)}^{[t]} = m_{i,p(n)}^{[h]} \left(1 - \frac{Net_{i,p(n)}}{C^2} \right); \tag{3.8}$$

d. Updating of the connections $w_{i,j(n)}$ through the variation $\Delta w_{i,j(n)}$, which again can be thought to be a trapping of the energy difference in Eq. (3.8),

$$\Delta w_{i,j(n)} = \sum_{p}^{M} \left(m_{i,p(n)}^{[h]} - m_{i,p(n)}^{[t]} \right) \left(1 - \frac{w_{i,j(n)}}{C} \right) m_{j,p(n)}^{[h]} \tag{3.9}$$

$$w_{i,j(n+1)} = w_{i,j(n)} + \alpha \cdot \Delta w_{i,j(n)} \tag{3.10}$$

Even a cursory comparison of (3.4) and (3.8) and (3.5), (3.6), (3.9) and (3.10), respectively, clearly shows how both steps of the signal transfer process are guided by the same (contraction) principle, and likewise for the two weight updating steps which we consider to be an energy entrapment principle.

Notice how the term $m_{i,p(n)}^{[h]}$ in (3.9) makes the change in the connection, $w_{i,j(n)}$, proportional to the quantity of "energy" liberated by node, $m_{i,p(n)}^{[h]}$, in favour of node $m_{i,p(n)}^{[t]}$. The whole learning process, which essentially consists of a progressive adjustment of the connections aimed at the global transformation of "energy", may be seen as a complex juxtaposition of phases of acceleration and deceleration of velocities of the learning signals (the additive updating terms $\Delta w_{i,j(n)}$ and $\Delta v_{i(n)}$) inside the ANN connection matrix. To get a clearer understanding of this feature of the Auto-CM learning mechanics, begin by considering its convergence condition,

$$\lim_{n \to \infty} v_{i(n)} = C. \tag{3.11}$$

Indeed, when $v_{i(n)} = C$, then $\Delta v_{i(n)} = 0$ (according to Eq. (3.5)), and $m_{i,p(n)}^{[h]} = 0$ (according to Eq. (3.4)) and, consequently, $\Delta w_{i,j(n)} = 0$ (from Eq. (3.9)), Auto-CM then converges.

There are, moreover, four variables that play a key role in the learning mechanics of Auto-CM. Specifically we have the following observations.

1. $\varepsilon_{i(n)}$ is the contraction factor of the first layer of Auto-CM weights,

$$\varepsilon_{i(n)} = 1 - \frac{v_{i(n)}}{C} \tag{3.12}$$

In fact, we can think about the contraction factor as the two first terms of a Taylor expansion of exponential function,

$$\varepsilon_{i(n)} = e^{-\frac{v_{i(n)}}{C}} \approx 1 - \frac{v_{i(n)}}{C} \tag{3.13}$$

or

$$\frac{v_{i(n)}}{C} \approx 1 - \varepsilon_{i(n)} = 1 - e^{-\frac{v_{i(n)}}{C}}. \tag{3.14}$$

It is clear if $v_{i(n)} \to C$ then the contraction factor $\varepsilon_{i(n)} \to 0$. As may be apparent from (3.4), the parameter C modulates the transmission of the input signal into the hidden layer by 'squeezing' it for given values of the connections; the actual extent of the squeeze is indeed controlled by the value of C, thereby explaining its interpretation as the contraction parameter. Clearly, the choice of C and the initialization of the connection weights must be such that the contraction factor is a number such that $1/C$ always falls within the $(0,1)$ range or $C \in (1,q)$ for q large and $(1/C)^n$ is decreasing at every processing cycle n, to become infinitesimal as n gets (infinitely) larger. Note, given the iteration at k, the contraction is $(1/C)^k$. So as long as $C > 1$, this will serve as a contraction factor.

2. $\eta_{i,j(n)}$ is, analogously, the contraction factor of the second layer of Auto-CM weights which, once again given the initialization choice, falls strictly within the unit interval,

$$\eta_{i,j(n)} = 1 - \frac{w_{i,j(n)}}{C} \tag{3.15}$$

We can think about $\eta_{i,j(n)}$ in a similar manner as we did for $\varepsilon_{i(n)}$, that is, as the two first terms of a Taylor expansion of exponential function,

$$\eta_{i,j(n)} = e^{-\frac{w_{i,j(n)}}{C}} \approx 1 - \frac{w_{i,j(n)}}{C} \tag{3.16}$$

or

$$\frac{w_{i,j(n)}}{C} \approx 1 - \eta_{i,j(n)} = 1 - e^{-\frac{w_{i,j(n)}}{C}}. \tag{3.17}$$

It is clear as $w_{i,j(n)} \to C$ then $\eta_{i,j(n)} \to 0$. As in the previous layer, the value of the contraction factor is modulated by the contraction parameter C.

3. $\varphi_{i(n)}$ is the difference between the hidden and the input nodes:

$$\varphi_{i(n)} = m_{i,p(n)}^{[s]} - m_{i,p(n)}^{[h]} \tag{3.18}$$

It is a real function of n, and it always takes positive, decreasing values in view of the contractive character of the signal transfer process.

4. $\lambda_{i(n)}$ is, likewise, the difference between the output and the hidden nodes:

$$\lambda_{i(n)} = m^{[h]}_{i,p(n)} - m^{[t]}_{i,p(n)} \tag{3.19}$$

It is, by the same token, a real function with positive values, decreasing in n.

We gain further insight into the Auto-CM learning mechanics by looking at the second step during the Auto-CM learning phase, $\Delta v_{i(n)}$ which describes a parabolic arc, always lying in the positive orthant. To see this, according to Eqs. (3.4) and (3.5),

$$\Delta v_{i(n)} = \sum_{p}^{M} \left(m^{[s]}_{i,p} - m^{[s]}_{i,p} \left(1 - \frac{v_{i(n)}}{C} \right) \right) \left(1 - \frac{v_{i(n)}}{C} \right) m^{[s]}_{i,p} = \sum_{p}^{M} \frac{v_{i(n)}}{C} \left(1 - \frac{v_{i,n}}{C} \right) \left(m^{[s]}_{i,p} \right)^2$$

$$= \left(\sum_{p}^{M} \left(m^{[s]}_{i,p} \right)^2 \right) \left(\frac{v_{i(n)}}{C} \right) \left(1 - \frac{v_{i(n)}}{C} \right). \tag{3.20}$$

Remembering how $\varepsilon_{i(n)}$ was defined, we can write $\frac{v_{i(n)}}{C} = 1 - \varepsilon_{i(n)}$ and so that (3.20), as a function of the contraction factor $\varepsilon_{i(n)}$, becomes

$$\Delta v_{i(n)} = \sum_{p}^{M} \left(1 - \varepsilon_{i(n)} \right) \varepsilon_{i(n)} \left(m^{[s]}_{i,p} \right)^2 = \left(\sum_{p}^{M} \left(m^{[s]}_{i,p} \right)^2 \right) \left(1 - \varepsilon_{i(n)} \right) \varepsilon_{i(n)} \tag{3.21}$$

Keeping in mind the definition of $\varepsilon_{i(n)}$, and letting the values of the input layer units decrease within the unit interval, one can easily check that the $\Delta v_{i(n)}$ parabolic arc (3.21) meets the following condition,

$$0 < \frac{\Delta v_{i(n)}}{\left(\sum_{p}^{M} \left(m^{[s]}_{i,p} \right)^2 \right)} < \varepsilon_{i(n)} \leq C\varepsilon_{i(n)} \tag{3.22}$$

Equation (3.22) tells us that the update term, $\Delta v_{i(n)}$, of the connections between the input and hidden layers will always be smaller than the update that $v_{i(n)}$ needs to reach up to C, that is,

$$v_{i(n+1)} = v_{i(n)} + \Delta v_{i(n)} = C - C\varepsilon_{i(n)} + \Delta v_{i(n)} \tag{3.23}$$

But from (3.22) we know that

$$\frac{\Delta v_{i(n)}}{\left(\sum_{p}^{M} \left(m^{[s]}_{i,p} \right)^2 \right)} - C\varepsilon_{i(n)} \leq 0 \tag{3.24}$$

which proves our claim. As a consequence, $v_{i(n)}$ will never exceed C. However, the convergence condition requires that $\lim_{n\to\infty} v_{i(n)} = C$, which in turn implies the following:

$$\lim_{n\to\infty} \varepsilon_{i(n)} = 0, \quad \lim_{n\to\infty} m^{[h]}_{i,p(n)} = 0, \quad \lim_{n\to\infty} m^{[t]}_{i,p(n)} = 0,$$

$$\lim_{n\to\infty} \varphi_{i(n)} = m^{[s]}_{i,p(n)}, \quad \lim_{n\to\infty} \lambda_{i(n)} = 0 \tag{3.25}$$

Notice that $v_{i,n} \leq C - C\varepsilon_{i,n}\left(1 - \sum_p^M \left(m^{[s]}_{i,p}\right)^2\right)$.

The contractive character of the signal transmission process at both of the Auto-CM's layer levels, combining Eqs. (3.4) and (3.8), allows us to write:

$$m^{[t]}_{i,p(n)} \leq m^{[h]}_{i,p(n)} \leq m^{[s]}_{i,p(n)} \tag{3.26}$$

and in particular, we can reformulate them as follows:

$$m^{[h]}_{i,p(n)} = m^{[s]}_{i,p(n)} \cdot \varepsilon_{i(n)} \tag{3.27}$$

and

$$m^{[t]}_{i,p(n)} = m^{[s]}_{i,p(n)} \varepsilon_{i(n)} \left(1 - \frac{Net_{i(n)}}{C^2}\right) \tag{3.28}$$

We are now in the position to clarify the relationship between $\Delta v_{i(n)}$ and $\Delta w_{i,j(n)}$. From Eqs. (3.4)–(3.8), we can stipulate:

$$m^{[h]}_{i,p(n)} = m^{[s]}_{i,p(n)} - \varphi_{i(n)} \tag{3.29}$$

where $\varphi_{i(n)}$ is a small positive real number, and

$$m^{[t]}_{i,p(n)} = m^{[h]}_{i,p(n)} - \lambda_{i(n)} \tag{3.30}$$

where $\lambda_{i(n)}$ is another positive real number. As already remarked, such positive numbers must come close to 0 as n increases. We can also write

$$m^{[t]}_{i,p(n)} = m^{[s]}_{i,p(n)} - \left(\varphi_{i(n)} + \lambda_{i(n)}\right). \tag{3.31}$$

At this point, we can easily reformulate Eq. (3.5) as:

$$\Delta v_{i(n)} = \sum_p^M \left(m^{[s]}_{i,p} - m^{[h]}_{i,p(n)}\right)\left(1 - \frac{v_{i(n)}}{C}\right) m^{[s]}_{i,p} = \sum_p^M \varphi_{i(n)} \cdot \varepsilon_{i(n)} m^{[s]}_{i,p} \tag{3.32}$$

And, likewise, we can reformulate Eq. (3.9) as:

$$\Delta w_{i,j(n)} = \sum_{p}^{M} \left(m_{i,p(n)}^{[h]} - m_{i,p(n)}^{[t]} \right) \left(1 - \frac{w_{i,j(n)}}{C} \right) m_{j,p(n)}^{[h]} = \sum_{p}^{M} \lambda_{i(n)} \cdot \eta_{i,j(n)} m_{j,p}^{[s]} \varepsilon_{j(n)} \quad (3.33)$$

noting that

$$\lim_{\varepsilon \to 0} \Delta w_{i,j(n)} = 0 \quad (3.34)$$

Plugging (3.33) into (3.11) and remembering the definition of the contraction factor $\eta_{i,j(n)}$ yields:

$$w_{i,j(n+1)} = C \left(1 - \eta_{i,j(n)} \right) + \alpha \sum_{p}^{M} \lambda_{i(n)} \cdot \eta_{i,j(n)} m_{i,p}^{[s]} \varepsilon_{i(n)} \quad (3.35)$$

Finally, from (3.35) and (3.11), we can conclude that:

$$\lim_{n \to \infty} w_{i,j(n)} = C \left(1 - \eta_{i,j(n)} \right) \quad (3.36)$$

$$\lim_{n \to \infty} \eta_{i,j,n} = \eta_{i,j(\infty)} \quad (3.37)$$

where $\eta_{i,j(\infty)}$ is a finite value, $0 < \eta_{i,j(\infty)} < 1$, and $\eta_{i,j(\infty)}$ is close to zero, for example, 0.001 is a usual choice. What we have presented means that

$$\lim_{n \to \infty} w_{i,j(n)} = \lim_{n \to \infty} C \left(1 - \eta_{i,j(\infty)} \right) = w_{i,j(\infty)} < C. \quad (3.38)$$

Note: the decay rate of $\varepsilon_{i(n)}$ is faster than $\eta_{i,j(n)}$. Therefore when $\varepsilon_{i(n)}$ approaches zero, $\eta_{i,j(n)}$ approaches to an asymptotic value $\eta_{i,j(\infty)}$ which is close to 0 though in general greater or a little greater than 0. At the same time $w_{i,j(n)}$ approaches its finite asymptotic value which is less than C, but close to C.

The learning mechanics of the Auto-CM, in a nutshell, boils down to the following. At the beginning of the training, the input and hidden units will be very similar (see Eq. (3.4)), and, consequently, $\Delta v_{i(n)}$ will be very small (see Eq. (3.32)), while for the same reason $\lambda_{i(n)}$ (see its definition above) at the beginning will be very big and $\Delta w_{i,j(n)}$ bigger than $\Delta v_{i(n)}$ (see Eqs. (3.32) and (3.33)). During the training, while $v_{i(n)}$ rapidly increases as the processing cycles n proceed, $m_{i,p(n)}^{[h]}$ decreases, and so do accordingly $\lambda_{i(n)}$ and $\varepsilon_{i(n)}$. Consequently, $\Delta w_{i,j(n)}$ rolls along a downward slope, whereas $\Delta v_{i(n)}$ slows down the pace of its increase. When $\lambda_{i(n)}$ becomes close to zero, this means that $m_{i,p(n)}^{[h]}$ is now only slightly bigger than $m_{i,p(n)}^{[t]}$ (see Eq. (3.30)). $\Delta v_{i(n)}$ is accordingly getting across the global maximum of the equation $\Delta v_{i(n)} = \left(\sum_{p}^{M} \left(m_{i,p}^{[s]} \right)^2 \right) \left(1 - \varepsilon_{i(n)} \right) \cdot \varepsilon_{i(n)}$, so once the critical point has been hit, $\Delta v_{i(n)}$ will in turn begin its descent toward zero.

3.3   The Auto-contractive Map: An Example

The above Auto-CM mechanics will be illustrated by means of an example dataset. Specifically, consider the following 3 bits dataset (Table 3.1):

Auto-CM, after 48 epochs of processing with C > 1, has completed its learning of the dataset with a RMSE = 0.0000. Denoting by **v** the three weights of the first layer, at the end of the training, the coordinates of **v** are given in Table 3.2.

Table 3.3 shows the weights of the second layer at the end of the training.

Table 3.4, reports the dynamics of the training of the weights of the first layer, and of the only three weights of the second layer that connect different nodes (w(1,2), w(1,3), w(2,3)).

It is apparent that, as expected, all weights monotonically converge. Consider, as a further useful illustration (Fig. 3.2), the graph of the dynamics of weights v(1,1) and w(1,2), and compare it with the graph showing their updating (i.e., the respective adaptations) during the training phase (Fig. 3.3).

We can clearly trace these graphs back to the dynamics described in the Eqs. (3.11)–(3.36), and to the simple reconstruction of the mechanics spelled out just above. We can make similar remarks by considering the graph of the history of the values taken on by the first hidden node and by the first output node (Fig. 3.4).

Table 3.1 3 Bits dataset

3 Bits	Var 1	Var 2	Var 3
Rec 1	0	0	0
Rec 2	0	0	1
Rec 3	0	1	0
Rec 4	0	1	1
Rec 5	1	0	0
Rec 6	1	0	1
Rec 7	1	1	0
Rec 8	1	1	1

Table 3.2 Weights of the first layer at the end of the training

v(1)	1.00
v(2)	1.00
v(3)	1.00

Table 3.3 Weights of the second layer at the end of the training

W(i,j)	Bit 1	Bit 2	Bit 3
Bit 1	0.95114	0.861754	0.861754
Bit 2	0.861754	0.95114	0.861754
Bit 3	0.861754	0.861754	0.95114

Table 3.4 Dynamics of the training

3bits	v(1)	v(2)	v(3)	w(1,2)	w(1,3)	w(2,3)
Epoch1	0.000161	0.000161	0.000161	0.370855	0.370856	0.370857
Epoch2	0.000259	0.000259	0.000259	0.533956	0.533957	0.533959
Epoch3	0.000418	0.000418	0.000418	0.627853	0.627855	0.627857
Epoch4	0.000672	0.000672	0.000672	0.689412	0.689414	0.689416
Epoch5	0.001083	0.001083	0.001083	0.733061	0.733064	0.733066
Epoch6	0.001742	0.001742	0.001742	0.765688	0.765692	0.765695
Epoch7	0.002803	0.002803	0.002803	0.791018	0.791022	0.791026
Epoch8	0.004508	0.004508	0.004508	0.811242	0.811248	0.811253
Epoch9	0.007242	0.007242	0.007242	0.827736	0.827743	0.827750
Epoch10	0.011617	0.011617	0.011617	0.841397	0.841407	0.841416
Epoch11	0.018589	0.018589	0.018589	0.852828	0.852841	0.852854
Epoch12	0.029631	0.029631	0.029631	0.862436	0.862454	0.862471
Epoch13	0.046947	0.046947	0.046947	0.870492	0.870515	0.870539
Epoch14	0.073678	0.073678	0.073678	0.877163	0.877195	0.877227
Epoch15	0.113961	0.113961	0.113961	0.882545	0.882589	0.882633
Epoch16	0.172487	0.172487	0.172487	0.886696	0.886753	0.886811
Epoch17	0.253100	0.253100	0.253100	0.889672	0.889746	0.889820
Epoch18	0.356198	0.356198	0.356198	0.891592	0.891682	0.891773
Epoch19	0.475946	0.475946	0.475946	0.892666	0.892770	0.892875
Epoch20	0.599992	0.599992	0.599992	0.893172	0.893285	0.893400
Epoch21	0.713680	0.713680	0.713680	0.893370	0.893487	0.893608
Epoch22	0.806393	0.806393	0.806393	0.893435	0.893554	0.893677
Epoch23	0.874853	0.874853	0.874853	0.893453	0.893573	0.893697
Epoch24	0.921696	0.921696	0.921696	0.893458	0.893578	0.893702
Epoch25	0.952067	0.952067	0.952067	0.893459	0.893579	0.893703
Epoch26	0.971068	0.971068	0.971068	0.893459	0.893579	0.893703
Epoch27	0.982688	0.982688	0.982688	0.893459	0.893579	0.893703
Epoch28	0.989697	0.989697	0.989697	0.893459	0.893579	0.893703
Epoch29	0.993887	0.993887	0.993887	0.893459	0.893579	0.893703
Epoch30	0.996380	0.996380	0.996380	0.893459	0.893579	0.893703
Epoch31	0.997859	0.997859	0.997859	0.893459	0.893579	0.893703
Epoch32	0.998735	0.998735	0.998735	0.893459	0.893579	0.893703
Epoch33	0.999252	0.999252	0.999252	0.893459	0.893579	0.893703
Epoch34	0.999558	0.999558	0.999558	0.893459	0.893579	0.893703
Epoch35	0.999739	0.999739	0.999739	0.893459	0.893579	0.893703

(continued)

Table 3.4 (continued)

3bits	v(1)	v(2)	v(3)	w(1,2)	w(1,3)	w(2,3)
Epoch36	0.999846	0.999846	0.999846	0.893459	0.893579	0.893703
Epoch37	0.999909	0.999909	0.999909	0.893459	0.893579	0.893703
Epoch38	0.999946	0.999946	0.999946	0.893459	0.893579	0.893703
Epoch39	0.999968	0.999968	0.999968	0.893459	0.893579	0.893703
Epoch40	0.999981	0.999981	0.999981	0.893459	0.893579	0.893703
Epoch41	0.999989	0.999989	0.999989	0.893459	0.893579	0.893703
Epoch42	0.999993	0.999993	0.999993	0.893459	0.893579	0.893703
Epoch43	0.999996	0.999996	0.999996	0.893459	0.893579	0.893703
Epoch44	0.999998	0.999998	0.999998	0.893459	0.893579	0.893703
Epoch45	0.999999	0.999999	0.999999	0.893459	0.893579	0.893703
Epoch46	0.999999	0.999999	0.999999	0.893459	0.893579	0.893703
Epoch47	0.999999	0.999999	0.999999	0.893459	0.893579	0.893703
Epoch48	1.000000	1.000000	1.000000	0.893459	0.893579	0.893703

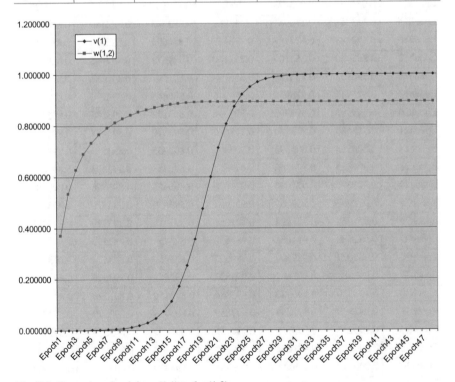

Fig. 3.2   Dynamics of weights v(1,1) and w(1,2)

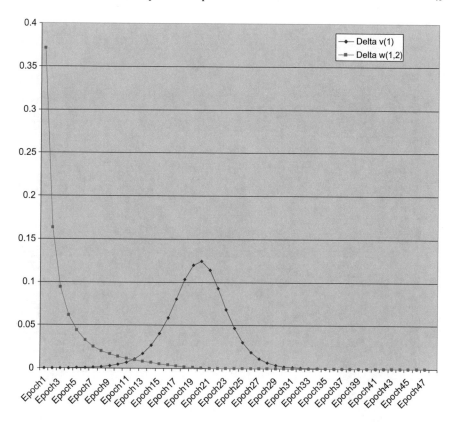

Fig. 3.3 Dynamics of updating of weights v(1,1) and w(1,2) during the training phase

Once again, the relaxation dynamics that has by now become typical of Auto-CM is apparent.

A further, important remark is that we can see the second layer connections of Auto-CM as the place where the energy liberated by the nodes, moving from one layer to another, is trapped. Figure 3.5, showing the dynamics of the contractive factors ε, η, and of the signal differences φ, λ, provides a clear illustration of this effect.

We can easily see from the Fig. 3.5 how, as n piles up, $n \rightarrow \infty$, energy is accumulated in the passage from the Input to the Hidden layer (the dynamics of φ) to be subsequently released in the passage from the Hidden to the Output layer (the dynamics of λ). Meanwhile, the contraction factors progressively die down.

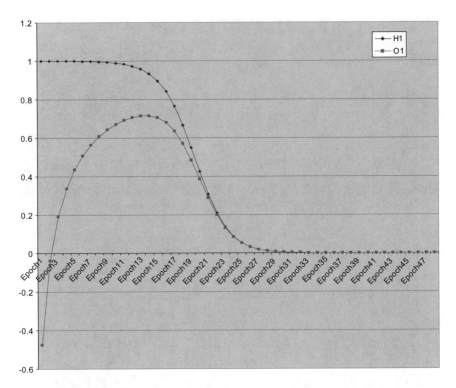

Fig. 3.4 Graph of the history of the values taken by the first hidden node and by the first output node

3.4   Experimenting with the Auto-contractive Map

Now that we have understood how the learning mechanics of the Auto-CM actually works, we are ready to explore its performance in carrying out tasks of interest. To this purpose, we now move our attention from the algorithmic structure of the Auto-CM to its actual behavior in relevant circumstances. In particular, we start by addressing the following basic issues:

a. How it behaves with respect to specific typologies of inputs;
b. How and whether it stabilizes its own output;
c. How its connections manage to stabilize.

To develop these points, we have chosen another toy problem whose nature fits particularly well in illustrating some of the qualifying characteristics and properties of this ANN, thereby allowing us to highlight them in an easy and simple way. Our input set consists of 9 patterns, each one composed of 121 nodes, which consist of sketchy pictures of human faces bearing a different expression (Fig. 3.6).

Given the structure of the Input, the Auto-CM has been constructed in the following way:

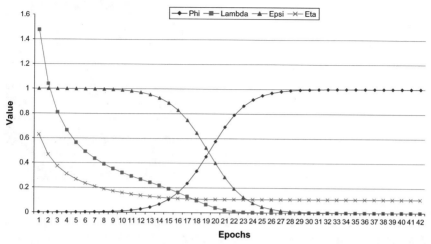

Fig. 3.5 Dynamics of the contractive factors

Fig. 3.6 Each face has been drawn within a matrix X whose cells can take only binary values (0 or 1). It is useful to point out that the Auto-CM has no information at all about the spatial (let alone the semantic) features of the Inputs, and simply considers the actual values of the matrix cells

- 121 Input nodes;
- 121 Hidden nodes;
- 121 Output nodes;
- 121 connections between Inputs and Hidden;
- 14,641 connections (i.e., 121×121) between the Hidden and Output units.

Fig. 3.7 Output for all of
the nine Input patterns after
two epochs

It is useful to point out that the Auto-CM has no information at all about the
spatial, let alone the semantic, features of the inputs, and simply considers the actual
values of the matrix cells.

All the 14,762 connections **v** and **w** have been initialized using the same value
(0.01). The signal transfer and the learning equations used here are the ones already
described above by Eqs. (3.4)–(3.10). The learning process has been carried out
presenting the nine patterns to the Auto-CM in a random way. The notion of epoch in
its more traditional meaning, namely, each epoch amounts to a complete presentation
to the ANN of all the training patterns in the Input set.

The ANN's performance during the training process may be divided into five
characteristic phases. In a **first phase** of the period of training, the output of all
the nine patterns tends to assume the value 1 for all the input nodes that belong to
the non-empty intersection of all the patterns, and the value 0 for all the rest of the
output nodes. This phase may be thought of as a phase of research of the **common
traits** among the whole set of the input patterns. An example of such an output is
the following (Fig. 3.7):

The **second phase**, triggers in each input vector a response that corresponds to the
union of all the patterns as the characteristic output. Here, the ANN may be regarded
as undertaking a phase of exploration and characterization of the overall variability
of the input patterns (Fig. 3.8):

The **third phase** input pattern is reproduced in the output vector by the Auto-CM
exactly as it is. This is clearly the phase of the sorting out of the specificity, that is,
of the *identification* of each item in the input set.

The **fourth phase** the output vector contains only all those entries of each specific
input pattern that do not lie in the common intersection of all patterns. In other words,
it selects the difference between the actual face and the 'intersection' face generated
at phase 1. This has to do with characterizing the differential traits that have to be
looked at after for the *discrimination* among the various patterns (Fig. 3.9):

The **last phase** input produces a null (zero) output. The ANN has acquired all the
necessary and useful elements for the pattern recognition task and has consequently

Fig. 3.8 Output response as the union of all input patterns after around 10 epochs

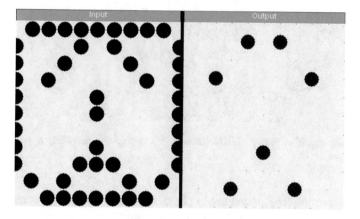

Fig. 3.9 Each pattern generates as output only its specific features, after around 100 epochs

ceased to learn, becoming unresponsive to the training stimuli as if it had 'lost interest' in them. This amounts to a *self-validation* of the Auto-CM, which thereby acknowledges that the learning process has come to an end.

This highly organized and systematic sequence of behavioural responses of the Auto-CM may seem surprising. Even more surprising is the outcome of the analysis of the structure of the 14,641 stabilized connections established between the hidden and the output layers of the ANN. In fact, such connections actually make up a single, global picture that can be thought to present fractal characteristics. The picture clearly draws out the 'union' face found in phase 2, that is, the one summarizing the overall variability of traits of all the faces included in the input set. But this time the 'big face' is no longer made of single pixels, but in turn of smaller faces, arranged in ways that encompass—in a highly structured, symmetrical way—all the above discussed learning phases. And the way in which the small faces are chosen to represent details of the big faces is highly pertinent. To all of the points of the 'big face' that correspond

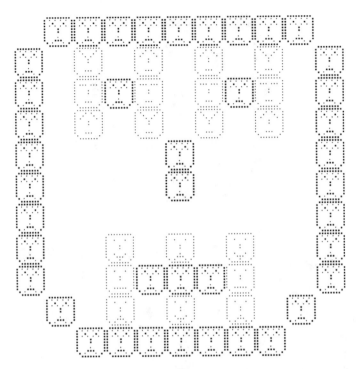

Fig. 3.10 The matrix of values of the Hidden-Output connections according to the B-operator where $n_R = n_C = 11$

to the 'intersection face' found out in phase 1 of the process, there correspond an 'intersection face'. The 'union face' is pictured only in the 'big face', but is never found in the smaller ones, in that it corresponds to the overall variability and it is pointless to assign it at any specific point of the 'big face'. The rest of the points of the 'union face' host faces that resemble the original input patterns, placed in locations that correspond to their typical differential traits, but no one is really identical to any of the actual input patterns. In particular, in the various 'small faces' one can find all the "expressions of the face" and all the "expressions of the eyes" occurring in the nine training patterns.

Specifically, if the "expression of the mouth" is equal to the expression of the mouth of any of the training patterns, then the "expression of the eyes" is the *union* of all the expressions of the eyes of the corresponding training patterns. Similarly, if the "expression of the eyes" is equal to the one of any training patterns, then the expression of the mouth is the *union* of all the expressions of the mouths of the corresponding training patterns. This ingenious 'cognitive' scheme ultimately may represent a *fractal immersion* of the N dimensional space of the input into the N^2 dimensional space associated to the weights matrix of the Auto-CM (Fig. 3.10).

The specificity of this particular set of inputs and of the related tasks notwithstanding, the insights developed here hold more generally and characterize Auto-CM's performance. In particular, it turns out that one of the Auto-CM's specialties is that of finding out, within a given set of input patterns, the global statistics of the associations among the underlying variables. Also the peculiarities of the matrix of the hidden-output connections have been reproduced in hundreds of tests with very diverse input classes of all kinds.

3.5 Auto-CM: A Theoretical Discussion

There are a few important peculiarities of Auto-CMs with respect to more familiar classes of ANNs that need special attention and careful reflection.

- Auto-CMs are able to learn also when starting from initializations where all connections are set at the same value, that is, they do not suffer the problem of the symmetric connections.
- During the training process, Auto-CMs always assign positive values to connections. In other words, Auto-CMs do not allow for inhibitory relations among nodes, but only for different strengths of excitatory connections.
- Auto-CMs can learn also in difficult conditions, namely, when the connections of the main diagonal of the second layer connection matrix are removed. In the context of this kind of learning process, Auto- CMs seem to reconstruct the relationship occurring between each couple of variables. Consequently, from an experimental point of view, it seems that the ranking of its connections matrix translates into the ranking of the joint probability of occurrence of each couple of variables.
- At the end of the training phase ($\Delta w_{i,j} = 0$), all the components of the weights vector $\mathbf{v}$ reach the same value:

$$\lim_{n \to \infty} v_{i(n)} = C. \tag{3.39}$$

Recall that it was mentioned that as $n \to \infty$, the output and hidden values approaches zero, $\lim_{n \to \infty} m^{[t]}_{n,p(n)} = 0$, and $\lim_{n \to \infty} m^{[h]}_{n,p(n)} = 0$, and consequently according Eq. (3.9) Δw approaches zero while $\mathbf{w}$ is less than C.

The matrix $\mathbf{w}$, then, represents the Auto-CM knowledge about the whole dataset. One can use the information embedded in the $\mathbf{w}$ matrix to compute in a natural way the joint probability of occurrence among variables:

$$P_{i,j} = \frac{w_{i,j}}{\sum_{k=1}^{N} w_{i,k}} \tag{3.40}$$

$$P\left(m^{[s]}_i\right) = \sum_{j}^{N} P_{i,j} = 1 \tag{3.41}$$

In fact,

$$\sum_i^N P_{i,j} = \sum_j^N \frac{w_{i,j}}{\sum_{k=1}^N w_{i,k}} = 1 \qquad (3.42)$$

The new matrix **p** can be read as the probability of transition from any state variable to any other, that is,

$$P\left(m_i^{[t]}|m_j^{[s]}\right) = P_{i,j} \qquad (3.43)$$

Alternatively, the matrix **w** may be transformed into a non-Euclidean distance metric (semi-metric), when we train the Auto-CM with the main diagonal of the **w** matrix fixed at value N.

Remark We have two different points of view in our interpretation of weights generating probabilities. One is from the point of view of rows and the other is of columns. We note that from this point of view, $P_{i,j}$ is not the same as $P_{j,i}$.

Now, if we consider C as a limit value for all the weights of the **w** matrix, we can write

$$d_{i,j} = C - w_{i,j} \qquad (3.44)$$

The new matrix **d** is again a squared non symmetric matrix, where the main diagonal entries are null (i.e., they represent the zero distance of each variable from itself), and where the off-diagonal entries represent "distances" between each couple of variables. We will expand on this interpretation next.

Contractive Factor We now discuss in more detail the "spatial" interpretation of the squared weights matrix of the Auto-CM. Consider each variable of the dataset as a vector made by all of its values. From this perspective, one can see the dynamic value of each connection between any two variables (the hidden and output nodes) as the "local velocity", $\Delta w_{i,j}$, of their mutual attraction caused by the degree of similarity of their vectors. We can extract from each weight of a trained Auto-CM this specific inverse contractive factor as

$$F_{i,j} = \left(1 - \frac{w_{i,j}}{C}\right)^{-1} \quad 1 \le F_{i,j} \le \infty. \qquad (3.44a)$$

Equation (3.44a) is the inverse of the contractive factor. Equation (3.44a) is interesting for 2 reasons.

1. It is the inverse of the contractive factor that controls the Auto-CM training. That is, in view of Eq. (3.11), each mono-connection v_i at the end of the training will reach the value C. In this case, the contractive factor will diverge because the two variables connected by the weight are indeed the same variable.
2. Considering instead Eq. (3.36), each weight, $w_{i,j}$, at the end the training will always be smaller than C. This means that the contractive factor for each weight

of the matrix will always stay bounded. To visualize this claim within our spatial reference space, notice that, in the case of the weight, $w_{i,j}$, the variable is of course connected with itself, but the same variable has also received the influences of the other variables (recall that the matrix **w** is a squared matrix where each variable is linked to the other), and consequently there will remain enough difference to prevent contractive collapse.

We are now in the position to calculate the contractive distance between each variable and the other, by suitably adjusting the original Euclidean distance by the specific contractive factor Eq. (3.44a). The Euclidean distance among the variables in the dataset is given by the following equation:

$$d_{i,j}^{[\text{Euclidean}]} = \sqrt{\sum_{k}^{R} \left(x_{i,k} - x_{j,k}\right)^2} \tag{3.45}$$

where X are the dataset values used as input, R = the number of the records of the assigned dataset; $x_{i,k}$ and $x_{j,k}$ = the ith value and the jth value of the variables in the kth record;

Whereas the Auto-CM distance matrix among the same variables is given by

$$d_{i,j}^{[\text{AutoCM}]} = \frac{d_{i,j}^{[\text{Euclidean}]}}{F_{i,j}} \tag{3.46}$$

Equation (3.46) makes it explicit how the similarity attraction effect deforms the original Euclidean embedding space and paves the way to a fruitful characterization and interpretation of the 'spatializing' properties of Auto-CM, to which we now turn. When the F factor is 1 then **w** = 0 and the Auto-CM distance between two variables is the same as the Euclidean distance (or better similarity).

Summary We have presented the underlying theory, topology and the algorithm of a new ANN, Auto Contractive Map (Auto-CM). The Auto-CM system reshapes the distances among variables or records of any dataset, considering their global vectorial similarities and consequently defining the specific "warped space" into which variables or records can be embedded. This new warped space into which the dataset is embedded is the one in which the relationships are extracted. The display of the dataset in this "warped space" is Chap. 4.

Reference

1. Buscema, M., C. Helgason, and E. Grossi. 2008. Auto contractive maps, H function and maximally regular graph: Theory and applications. In *Special session on "Artificial Adaptive Systems in Medicine: Applications in the Real World, NAFIPS 2008 (IEEE)"*, New York, May 19–22, 2008.

General References

Buscema, M. (ed.). 2007. Squashing Theory and Contractive Map Network. *Semeion Technical Paper #32*, Rome.

Buscema, M. 2007. A novel adapting mapping method for emergent properties discovery in data bases: Experience in medical field. In *2007 IEEE International Conference on Systems, Man and Cybernetics (SMC 2007)*, Montreal, Canada, Octobre 7–10, 2007.

Buscema, M., and E. Grossi. 2008. The Semantic Connectivity Map: An Adapting Self-organizing Knowledge Discovery Method in Data Bases. Experience in Gastro-oesophageal Reflux Disease. *International Journal of Data Mining and Bioinformatics* 2 (4).

Buscema, M., E. Grossi, D. Snowdon, and P. Antuono. 2008. Auto-contractive Maps: An Artificial Adaptive System for Data Mining. An Application to Alzheimer Disease. *Current Alzheimer Research* 5: 481–498.

Licastro, F., E. Porcellini, M. Chiappelli, P. Forti, M. Buscema, et al. 2010. Multivariable Network Associated with Cognitive Decline and Dementia. *International Neurobiology of Aging* 1 (2): 257–269.

Buscema, M., and E. Grossi (eds.). 2009. *Artificial Adaptive Systems in Medicine*, 25–47. Bentham e-books.

Buscema, Massimo, and Pier L. Sacco. 2010. Auto-contractive Maps, the H Function, and the Maximally Regular Graph (MRG): A New Methodology for Data Mining. In *Applications of Mathematics in Models, Artificial Neural Networks and Arts*, ed. V. Capecchi, et al. Berlin: Springer.

Grossi, Enzo, Giorgio Tavano Blessi, Pier Luigi Sacco, and Massimo Buscema. 2011. The Interaction Between Culture, Health and Psychological Well-Being: Data Mining from the Italian Culture and Well-Being Project. *Journal of Happiness Studies* (Springer).

Licastro, Federico, Elisa Porcellini, Paola Forti, Massimo Buscema, Ilaria Carbone, Giovanni Ravaglia, and Enzo Grossi. 2010. Multi Factorial Interactions in the Pathogenesis Pathway of Alzheimer's Disease: A New Risk Charts for Prevention of Dementia. *Immunity & Ageing* 7 (Suppl 1): S4.

Buscema, M., F. Newman, E. Grossi, and W. Tastle. 2010. Application of Adaptive Systems Methodology to Radiotherapy. In *NAFIP*, Toronto, Canada, July 12–14, 2010.

Eller-Vainicher, C., V.V. Zhukouskaya, Y.V. Tolkachev, S.S. Koritko, E. Cairoli, E. Grossi, P. Beck-Peccoz, I. Chiodini, and A.P. Shepelkevich. 2011. Low Bone Mineral Density and Its Predictors in Type 1 Diabetic Patients Evaluated by the Classic Statistics and Artificial Neural Network Analysis. *Diabetes Care* 1–6.

Gomiero, T., L. Croce, E. Grossi, L. De Vreese, M. Buscema, U. Mantesso, and E. De Bastiani. 2011. A Short Version of SIS (Support Intensity Scale): The Utility of the Application of Artificial Adaptive Systems. US-China Education Review A 2: 196–207.

Buscema, M., S. Penco, and E. Grossi. 2012. A Novel Mathematical Approach to Define the Genes/SNPs Conferring Risk or Protection in Sporadic Amyotrophic Lateral Sclerosis Based on Auto Contractive Map Neural Networks and Graph Theory. *Neurology Research International* 2012: 13, Article ID 478560.

Grossi, E., A. Compare, and M. Buscema. 2012. The Concept of Individual Semantic Maps in Clinical Psychology: A Feasibility Study on a New Paradigm. *Quality & Quantity International Journal of Methodology*, August 04th, 2012.

Coppedè, F., E. Grossi, M. Buscema, and L. Migliore. 2013. Application of Artificial Neural Networks to Investigate One-Carbon Metabolism in Alzheimer's Disease and Healthy Matched Individuals. *PLOS ONE* 8 (8): e74012, 1–11.

Street, M.E., M. Buscema, A. Smerieri, L. Montanini, and E. Grossi. 2013. Artificial Neural Networks, and Evolutionary Algorithms as a Systems Biology Approach to a Data-base on Fetal Growth Restriction. *Progress in Biophysics and Molecular Biology*, 1–6.

Compare, A., E. Grossi, M. Buscema, C. Zarbo, X. Mao, F. Faletra, E. Pasotti, T. Moccetti, P.M.C. Mommersteeg, and A. Auricchio. 2013. Combining Personality Traits with Traditional Risk Factors for Coronary Stenosis: An Artificial Neural Networks Solution in Patients with Computed

Tomography Detected Coronary Artery Disease. *Cardiovascular Psychiatry and Neurology* 2013: 9, Article ID 814967 (Hindawi Publishing Corporation).

Buscema, M., V. Consonni, D. Ballabio, A. Mauri, G. Massini, M. Breda, and R. Todeschini. 2014. K-CM: A New Artificial Neural Network. Application to Supervised Pattern Recognition. *Chemometrics and Intelligent Laboratory Systems* 138: 110–119.

Buscema, M., G. Massini, and G. Maurelli. 2014. Artificial Neural Networks: An Overview and Their Use in the Analysis of the AMPHORA-3 Dataset. *Substance Use & Misuse*, Early Online: 1–14.

Gironi, M., B. Borgiani, E. Farina, E. Mariani, C. Cursano, M. Alberoni, R. Nemni, G. Comi, M. Buscema, R. Furlan, and Enzo Grossi. 2015. A Global Immune Deficit in Alzheimer's Disease and Mild Cognitive Impairment Disclosed by a Novel Data Mining Process. *Journal of Alzheimer's Disease* 43: 1199–1213.

Drenos, F., E. Grossi, M. Buscema, and S.E. Humphries. 2015. Networks in Coronary Heart Disease Genetics as a Step Towards Systems Epidemiology. *PLoS ONE* 10 (5): e0125876. https://doi.org/10.1371/journal.pone.0125876.

Coppedè, F., E. Grossi, A. Lopomo, R. Spisni, M. Buscema, and Lucia Migliore. 2015. Application of Artificial Neural Networks to Link Genetic and Environmental Factors to DNA Methylation in Colorectal Cancer. *Epigenomics* 7 (2): 175–186.

Narzisi, A., F. Muratori, M. Buscema, S. Calderoni, and E. Grossi. 2015. Outcome Predictors in Autism Spectrum Disorders Preschoolers Undergoing Treatment as Usual: Insights from an Observational Study Using Artificial Neural Networks. *Neuropsychiatric Disease and Treatment* 11: 1587–1599.

Buscema, M., E. Grossi, L. Montanini, M.E. Street. 2015. Data Mining of Determinants of Intrauterine Growth Retardation Revisited Using Novel Algorithms Generating Semantic Maps and Prototypical Discriminating Variable Profiles. *PLoS ONE* 10 (7): e0126020. https://doi.org/10.1371/journal.

Buscema, Paolo Massimo, Lara Gitto, Simone Russo, Andrea Marcellusi, Federico Fiori, Guido Maurelli, Giulia Massini, and Francesco Saverio Mennini. 2016. The Perception of Corruption in Health: AutoCM Methods for an International Comparison. *Quality & Quantity*. https://doi.org/10.1007/s11135-016-0315-4 (Springer).

Buscema, Massimo, Masoud Asadi-Zeydabadi, Weldon Lodwick, and Marco Breda. 2016. The H0 Function, a New Index for Detecting Structural/Topological Complexity Information in Undirected Graphs. *Physica A* 447: 355–378.

Buscema, Paolo Massimo, Guido Maurelli, Francesco Saverio Mennini, et al. 2016. Artificial Neural Networks and Their Potentialities in Analyzing Budget Health Data: An Application for Italy of What-If Theory. *Quality & Quantity*. https://doi.org/10.1007/s11135-016-0329-y (Springer).

Buscema, Massimo, and Pier Luigi Sacco. 2016. MST Fitness Index and Implicit Data Narratives: A Comparative Test on Alternative Unsupervised Algorithms. *Physica A* 461: 726–746.

Ferilli, Guido, Pier Luigi Sacco, Emanuele Teti, and Massimo Buscema. 2016. Top Corporate Brands and the Global Structure of Country Brand Positioning: An AutoCM ANN Approach. *Expert Systems With Applications* 66: 62–75.

Caffarra, Paolo, Simona Gardini, Francesca Dieci, et al. 2013. The Qualitative Scoring MMSE Pentagon Test (QSPT): A New Method for Differentiating Dementia with Lewy Body from Alzheimer's Disease. *Behavioural Neurology* 27: 213–220. https://doi.org/10.3233/ben-120319.

Campisi, Giuseppina, Martina Chiappelli, Massimo De Martinis, Vito Franco, Lia Ginaldi, Rosario Guiglia, Federico Licastro, and Domenico Lio. 2009. Pathophysiology of Age-Related Diseases. *Immunity & Ageing* 6: 12. https://doi.org/10.1186/1742-4933-6-12.

Coppedè, Fabio, Enzo Grossi, Francesca Migheli, and Lucia Migliore. 2010. Polymorphisms in Folate-Metabolizing Genes, Chromosome Damage, and Risk of Down Syndrome in Italian Women: Identification of Key Factors Using Artificial Neural Networks. *BMC Medical Genomics* 3: 42.

De Benedetti, Stefano, Giorgio Lucchini, Alessandro Marocchi, Silvana Penco, Christian Lunetta, Stefania Iametti*, Elisabetta Gianazza, and Francesco Bonomi. 2015. Serum Metal Evaluation in a Small Cohort of Amyotrophic Lateral Sclerosis Patients Reveals High Levels of Thiophylic Species. *Peptidomics* 2: 29–34.

di Ludovico, Alessandro. 2008. Experimental approaches to glyptic art using artificial neural networks. An investigation into the Ur III iconological context. In *Proceedings of the 36th CAA Conference*, Budapest, April 2–6, 2008.

Di Ludovico, A., and G. Pieri. 2011. Artificial Neural Networks and Ancient Artefacts: Justifications for a Multiform Integrated Approach Using PST and Auto-CM Models. *Archeologia e Calcolatori* 22: 99–128.

Gallucci, Maurizio, Pierpaolo Spagnolo, and Maria Aricò. 2016. Predictors of Response to Cholinesterase Inhibitors Treatment of Alzheimer's Disease: Date Mining from the TREDEM Registry. Journal of Alzheimer's Disease 50: 969–979. https://doi.org/10.3233/JAD-150747.

Gironi, M., et al. 2013. A Novel Data Mining System Points Out Hidden Relationships Between Immunological Markers in Multiple Sclerosis. *Immunity & Ageing* 10: 1.

Grossi, E., S. Cazzaniga, S. Crotti, et al. 2014. The Constellation of Dietary Factors in Adolescent Acne: A Semantic Connectivity Map Approach. *Journal of the European Academy of Dermatology and Venereology*, December 2014. https://doi.org/10.1111/jdv.12878.

Knibbe, Ronald A., Mieke Derickx, Allaman Allamani, and Giulia Massini. 2014. Alcohol Consumption and Its Related Harms in the Netherlands Since 1960: Relationships with Planned and Unplanned Factors. *Substance Use & Misuse*, Early Online: 1–12.

Smerieri, Arianna, Chiara Testa, Pietro Lazzeroni, et al. 2015. Di-(2-Ethylhexyl) Phthalate Metabolites in Urine Show Age-Related Changes and Associations with Adiposity and Parameters of Insulin Sensitivity in Childhood. *PLoS ONE* 10 (2): e0117831. https://doi.org/10.1371/journal.pone.0117831.

Street, Maria E., Enzo Grossi, Cecilia Volta, Elena Faleschini, and Sergio Bernasconi. 2008. Placental Determinants of Fetal Growth: Identification of Key Factors in the Insulin-Like Growth Factor and Cytokine Systems Using Artificial Neural Networks. *BMC Pediatrics* 8: 24. https://doi.org/10.1186/1471-2431-8-24.

Chapter 4
Visualization of Auto-CM Output

Abstract One of the most powerful aspects of our approach to neural networks is not only the development of the Auto-CM neural network but the visualization of its results. In this chapter we look at two visualization approaches—the Minimal Spanning Tree (MST) and the Maximal Regular Graph (MRG). The resultant from Auto-CM is a matrix of weights. This weight matrix naturally fits into a graph theoretic framework since the weights connecting the nodes will be viewed as edges and the weights as the weights on these edges.

4.1 Introduction

The final output $w_{i,j}$ from the application of Auto-CM to a problem becomes even more useful when we are able to visualize it in terms of a graph. The transformation of the weights into a graph provides a visualization of the relationships that exists among the elements/variables (nodes). The complexity and connections among the variables can be clearly seen as we will show in Sect. 4.2.

The way we obtain a graph of the relationships is through Eq. (4.1) below that transforms the square weights matrix of Auto-CM into a square and a symmetric matrix of distances among nodes. Each distance between a pair of nodes may therefore be regarded as the weights on edges between these pair of nodes so that it becomes a graph-theoretic representation. In this way, the matrix $\mathbf{D} = (d_{i,j})$ of Eq. (4.1) itself may be analyzed using graph theory. The transformation we use is

$$d_{i,j} = C - \frac{w_{i,j} + w_{j,i}}{2} \qquad (4.1)$$

where C (an input to the Auto-CM algorithm, the reciprocal 1/C being the contractor) is the Auto-CM constant and $w_{i,j}$ is the final weight between ith and jth vertices. Note that we use $\mathbf{d}$ instead of $\mathbf{w}$ in order to transform weights into symmetric distances.

One application of a graph theoretical representation is to show the relationship among pairs of entities. Recall that a graph consists of a set of vertices, and a set of edges, where an edge is an object that connects two vertices in the graph. More precisely, a graph is a pair (V, E), where V is a finite set and E is a binary relation on V,

© Springer International Publishing AG, part of Springer Nature 2018
P. M. Buscema et al., *Artificial Adaptive Systems Using Auto Contractive Maps*, Studies in Systems, Decision and Control 131, https://doi.org/10.1007/978-3-319-75049-1_4

Table 4.1 Adjacency matrix
of a distance matrix

	A	B	C	D	...	Z
A	0	1	1	1	1	1
B	1	0	1	1	1	1
C	1	1	0	1	1	1
D	1	1	1	0	1	1
...	1	1	1	1	0	1
Z	1	1	1	1	1	0

to which it is possible to associate scalar values, in this case, the distances $d_{i,j}$. V is called a vertex set whose elements are called vertices. E is a collection of edges, where an edge is a pair (u, v) with u, v belonging to V. In a directed graph, edges are ordered pairs, connecting a source vertex to a target vertex. In an undirected graph, edges are not ordered pairs and they connect the two vertices in both directions, hence in an undirected graph (u, v) and (v, u) are two ways of writing the same edge.

The graph theoretic representation is not constrained by any *a priori* semantic restriction, that is, it does not say what a vertex or edge actually represents. They could be cities with connecting roads, or web-pages with hyperlinks, and so on. These semantic details are irrelevant to determine the graph structure and properties. The only thing that matters is that a specific graph may be taken as a proper representation of the phenomenon under study, to justify attention on that particular mathematical object.

An adjacency-matrix representation of a graph is a 2-dimensional $N_V \times N_V$ array, where N_V is the number of vertices and rows represent the list of vertices and columns represent edges among vertices. To each element in the array is assigned a Boolean value that indicates whether the edge (u, v) is in the graph.

A distance matrix among V vertices represents an undirected graph, where each vertex is linked with all the others except itself. Consider the following matrix (Table 4.1).

4.2 Minimal Spanning Tree

The Minimum Spanning Tree (MST) problem is defined as follows. Find an acyclic subset T of E that connects all of the vertices in the graph whose total weight, that is, the total distance in our case, is minimized. The total weight our case is given by

$$d(T) = \sum_{i=0}^{N_V-1} \sum_{j=i+1}^{N_V} d_{i,j}, \; \forall d_{i,j}. \tag{4.2}$$

T is called a spanning tree, and the MST is the T whose weighted sum of edges attains the minimum value

$$MST = Min\{d\,(T_k)\}.\tag{4.3}$$

Given an undirected graph G, representing a matrix of distances **d**, with V vertices, completely linked to each other, the total number of their edges (E) is

$$E = \frac{N_V\,(N_V - 1)}{2}.\tag{4.4}$$

and the number of its possible spanning trees is:

$$T = N_V^{N_V - 2}.\tag{4.5}$$

Kruskal (1956) found an algorithm to determine the MST of any undirected graph in a quadratic number of steps for the worst case [1, 2]. Obviously, the Kruskal algorithm generates one of the possible MSTs, not all of them. In fact, in a weighted graph more than one MST is possible.

The MST can be considered to represent the *energy minimization* state of a structure from a conceptual point of view. In fact, if we consider the basic components of a structure as vertices of a graph and the strength of the association among them as the weight of each edge, linking a pair of vertices, the MST represents the minimum of energy needed so that all the elements of the structure preserve the integrity of structure. In a closed system that occurs in nature, all the components tend to minimize the overall energy. So the MST, in specific situations, can represent the most probable state to which a system tends to become [3]. This is the reason why MST is so useful and is applied to many different fields [4–6].

The MST of an undirected graph considers each edge of the graph to be weighted [15, 16]. For Auto-CM Eq. (4.1) shows a way to obtain the weight each edge from the nodes, which are the variables of a dataset, where the weights are the final outputs that result after training. The MST applied to a specific set of weights of an undirected graph can have different degrees in their representations. What Eq. (4.1) measures can be considered to be the "compactness" of a graph. The smaller the MST weight (d_{ij}) is, the more "compact" this portion of the graph is, that is, the connection distance is "shorter". The more compact, the more the relevant the information of the graph weights matrix is well represented. To define the compactness of MST from numerical values is quite simple. As mentioned above, we define N_V to be the number of the nodes of a graph, $Q(w)$ is the sorting of the weight matrix of the graph from the larger value of the weights to the smaller values, and S as the number of the weights to be skipped in order to add the weakest or weights with the smallest values in the MST obtained from the $Q(w)$ vector. Then Eq. (4.6) below defines what we call the degree of compactness of the Auto-CM MST.

$$MST\,(Compactness) = \frac{N_V - 1}{(N_V - 1) + S};\tag{4.6}$$

If $S = 0$, the *MST (Compactness)* $= 1$, corresponding to the maximum value, denoted *Max (MST (Compactness))*.

If $S = \frac{(N_V - 1)(N_V - 2)}{2}$, then *MST (Compactness)* $= \frac{2}{N_V + 1}$ which is the minimum. We denote it *Min (MST (Compactness))*.

We remind the reader that S is the number of the weights to be skipped in order to add the weakest or weights with the smallest values in the MST obtained from the $Q(w)$ vector. This measure of compactness can be used with any kind of Auto-Associative ANN or any kind of Linear Auto-Association to generate a weight matrix among the variables of an assigned dataset.

Remark It is hard to train a two-layer Auto-Associative Back Propagation ANN with the main diagonal weights fixed (to avoid auto-correlation problems). In most cases, the Root Mean Square Error (RMSE) stops decreasing after a few epochs. This is especially true when the orthogonality of the records (the dot product between two records considered as a row vector is zero) is relatively high. This is frequently the case when we calculate the distance among the records of the dataset. When we need to calculate the distance (as is the case for getting the MST), it is necessary to train the transposed matrix of the dataset. It should be noted that if Linear Auto-Association is used for this purpose, all of the nonlinear associations among variables are lost. Therefore, Auto-CM seems to be the best choice to date to compute a complete nonlinear matrix of weights among variables or among records of any dataset.

The MST is often used as a filter on the correlations matrix between objects to highlight their most significant relationships [1, 2]. The MST is what we might call the nervous system, or minimal structure system, of any dataset. Indeed, adding up the connections predicted by the MST, if we consider the links as energy, we obtain the total energy of the system. The MST selects only those connections that minimize this energy, that is, only the links that are really necessary to maintain a cohesive system. Consequently, all links included in the MST are critical, but on the contrary, not every fundamental link of the data set is necessarily present in the MST. This limit is inherent to the very nature of the MST. Each connection that originates in an internal cycle in the graph is eliminated, regardless of its strength and significance which is part of the process of MST. In order to solve this problem and to better capture the inherent complexity of a dataset, it is necessary to add more connections to the MST on the basis of two criteria:

- New connections must be relevant from a quantitative point of view;
- New connections must be able to generate new cyclic microstructures that highlight the *regularities* in the data set from a qualitative point of view.

Consequently, our Auto-CM MST is transformed into an undirected cyclic graph [see Eq. (4.1)]. Because of the new cycles that are present, the graph now represents a dynamic system with a temporal dimension. This is the reason why this new graph should provide information not only on the structure but also on the functions of the dataset variables.

A new type of graph, constructed from the Auto-CM MST we call the Maximally Regular Graph (MRG) which discussed in next section, it is necessary to proceed as follows:

- Take the MST structure as a starting point of the new graph;
- Consider the ordered list of connections that were skipped as the MST was being generated;
- Estimate the function H_0 of the new graph which is obtained each time a new connection is added to the basic MST structure, in order to monitor the variation in the complexity of these graphs at every step.

4.2.1 The Maximally Regular Graph

We define Maximally Regular Graph (MRG) as the graph whose function H_0 (defined below) reaches the highest value among all the graphs generated by adding to the original MST the most relevant dataset connections. The concept and the algorithm to calculate the MRG was defined in [2, 3]. The Auto-CM MRG is constructed as follows:

$$H_i = f\left(G\left(A_i, N\right)\right) \qquad (4.7)$$

In Eq. (4.7) the generic function G where f is the identity function for a graph of A_i nodes and N arcs is [4]

$$H_i = \frac{\mu_i \cdot \varphi_i - 1}{A_i}. \qquad (4.8)$$

where μ_i and φ_i are quantities generated during the execution by the algorithm and related to the graph complexity. In particular μ_i measures the mean gradient of the graph while φ_i measures the dynamics of the gradient changes during the pruning process. Thus, Eq. (4.8) is a complex ratio between the mean gradient and the dynamics of this gradient, from one side, and the global graph connectivity from the other side. Calculation of the H_i function represents the complexity of the MST. Here H_0 will have a special significance and a discussion of why H_0 captures the complexity of the graph can be found in [4]. We denote

$$MRG = Max\{H_i\} \qquad (4.9)$$

Thus, the MRG is the graph with the highest value of H_i. The indices for the H function are:

$$i \in [0, 1, \ldots, R]. \qquad (4.10)$$

The index for the number of graph arcs is

$$p \in [N - 1, N, N + 1, \ldots, N - 1 + R] \, . \tag{4.11}$$

The number of the skipped arcs during the MST generation is

$$R \in \left[0, 1, \ldots, \frac{(N - 1)(N - 2)}{2} \right] . \tag{4.12}$$

The number "R" of Eq. (4.12) is a key variable in the calculation of the MRG. R could, in fact, be zero when the MST calculation does not include skipping any relevant connections. In this case, there is no MRG for that data set. Furthermore, R ensures that the last and consequently the weakest link added to generate the MRG is always more relevant than the weakest MST connection. The MRG, finally, generates the graph that presents the greatest number of regular microstructures that use the most important connections of the data set. The higher the value of the function H to generate the MRG, the more significant is the cyclic microstructure predicted by its graph.

4.3   Auto-CM and Maximally Regular Graph

The MST represents the nervous system of any dataset. In fact, the summation of the strength of the connections among all the variables can be thought to represent the total energy of that system. The MST selects only the connections **that minimize this energy**. Consequently, all the links shown by MST are fundamental, but not every fundamental link of the dataset is shown by MST.

Such limit is intrinsic to the nature of MST itself. Every link able to generate a cycle into the graph is eliminated, however its strength. To avoid this limit and to explain better the intrinsic complexity of a dataset, it is necessary to add more links to the graph according to two criteria:

- The new links have to be **relevant** from a quantitative point of view;
- The new links have to be able to generate new **cyclic regular microstructures**, from a qualitative point of view.

Consequently, the MST graph is transformed into an undirected graph with cycles. Because of the cycles, the new graph can be considered to be dynamic system, involving in its structure a **time** dimension. This is the reason why this new graph should provide information not only about the structure but also about the **functions** of the variables of the dataset.

To build this new graph we need to proceed in the following way:

- Assume the MST structure as a starting point of the new graph;
- Consider the sorted list of the connections skipped during the MST generation;
- Estimate the H Function of the new graph each time we add a new connection to the MST structure, to monitor the variation of the complexity of the new graph at every step.

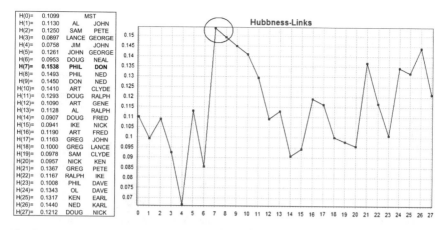

H(0)=	0.1099	MST	
H(1)=	0.1130	AL	JOHN
H(2)=	0.1250	SAM	PETE
H(3)=	0.0897	LANCE	GEORGE
H(4)=	0.0758	JIM	JOHN
H(5)=	0.1261	JOHN	GEORGE
H(6)=	0.0953	DOUG	NEAL
H(7)=	**0.1538**	**PHIL**	**DON**
H(8)=	0.1493	PHIL	NED
H(9)=	0.1450	DON	NED
H(10)=	0.1410	ART	CLYDE
H(11)=	0.1293	DOUG	RALPH
H(12)=	0.1090	ART	GENE
H(13)=	0.1128	AL	RALPH
H(14)=	0.0907	DOUG	FRED
H(15)=	0.0941	IKE	NICK
H(16)=	0.1190	ART	FRED
H(17)=	0.1163	GREG	JOHN
H(18)=	0.1000	GREG	LANCE
H(19)=	0.0978	SAM	CLYDE
H(20)=	0.0957	NICK	KEN
H(21)=	0.1367	GREG	PETE
H(22)=	0.1167	RALPH	IKE
H(23)=	0.1008	PHIL	DAVE
H(24)=	0.1343	OL	DAVE
H(25)=	0.1317	KEN	EARL
H(26)=	0.1440	NED	KARL
H(27)=	0.1212	DOUG	NICK

Fig. 4.1 The MRG Hubness of the Gang dataset

So, we have named MRG the graph whose **H Function** is the highest, among all the graphs generated adding to the original MST the new connections skipped before to complete the MST itself. The concept and the algorithm to calculate the MRG is found [2, 3].

4.3.1   An Example of the Maximally Regular Graph

Let us consider the Gang dataset (Table 4.2) to compute its MRG (Fig. 4.1). This is a small database of 27 records which describe the number of gang members in the "Jets and Sharks," taken from the musical West Side Story, and five variables that we use to characterize each record (name of gang member, gang name, age range, level of education, marital status and occupations).

In this example, the *H Function* reaches its peak when the seventh of the connections skipped during the MST generation is added back. So the MRG needs seven extra connections to be added to the MST and, consequently, the *H Function* value is almost 50% higher with respect to its value at the original MST (H(0) = 10.99 vs. H(7) = 15.38). It is then not surprising that the structure of the two graphs turns out very different (Figs. 4.2 and 4.3).

The MRG carries a richer amount of information than the MST. The boundary between the two Gangs, that is, between Jets and Sharks members, is now represented no longer by a couple of subjects, but rather by a cycle of four subjects: Neal and Ken are Sharks, while Doug and Mike are Jets. So, looking at the MRG, the edges between Jets and Sharks seem apparently to be fuzzy and negotiable, and in particular less clear-cut than in the MST case. But this appearance is very misleading. In fact, the four subjects lying on the border are all outliers in their respective gangs: By placing them all on the border, this feature becomes much clearer and provides a

Table 4.2 The Gang dataset

Name	Gang	Age	Education	Status	Occupations
ART	Jets	40	Junior School	Single	Pusher
AL	Jets	30	Junior School	Married	Burglar
SAM	Jets	20	College	Single	Bookie
CLYDE	Jets	40	Junior School	Single	Bookie
MIKE	Jets	30	Junior School	Single	Bookie
JIM	Jets	20	Junior School	Divorced	Burglar
GREG	Jets	20	High School	Married	Pusher
JOHN	Jets	20	Junior School	Married	Burglar
DOUG	Jets	30	High School	Single	Bookie
LANCE	Jets	20	Junior School	Married	Burglar
GEORGE	Jets	20	Junior School	Divorced	Burglar
PETE	Jets	20	High School	Single	Bookie
ERED	Jets	20	High School	Single	Pusher
GENE	Jets	20	College	Single	Pusher
RALPH	Jets	30	Junior School	Single	Pusher
PHIL	Sharks	30	College	Married	Pusher
IKE	Sharks	30	Junior School	Single	Bookie
NICK	Sharks	30	High School	Single	Pusher
DON	Sharks	30	College	Married	Burglar
NED	Sharks	30	College	Married	Bookie
KARL	Sharks	40	High School	Married	Bookie
KEN	Sharks	20	High School	Single	Burglar
EARL	Sharks	40	High School	Married	Burglar
RICK	Sharks	30	High School	Divorced	Burglar
OL	Sharks	30	College	Married	Pusher
NEAL	Sharks	30	High School	Single	Bookie
DAVE	Sharks	30	High School	Divorced	Pusher

better insight into what makes a 'typical' Jet or Shark. Furthermore, Al, a member of the Jets Gang, is placed at the top of an autonomous circuit of links among four Jets members, as he is the head of a new virtual gang hidden into the Jets Gang. Looking at the MRG, moreover, we also receive more information about the internal structure of the two Gangs: because of the bigger number of cycles, the Jets Gang reveals itself as more complex and articulated than the Sharks Gang. Finally, the cycle including Don, Ol, and Phil represents a prototype of the Sharks member, whose features are very different from Jets subjects. In the same way, Jets show two different prototypes; the first is represented by the cycle including Gene, Sam, Fred and Pete; the second by the cycle including John, George, Lance and Jim.

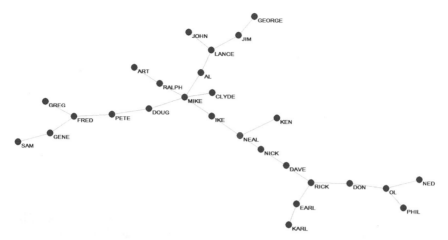

Fig. 4.2 MST of the Gang dataset

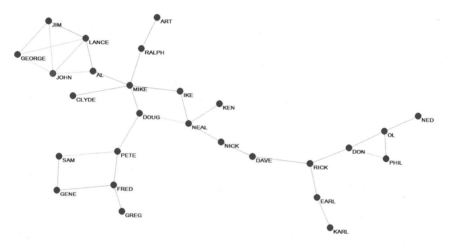

Fig. 4.3 MRG of the Gang dataset

Compared to the MST, therefore, the MRG adds those extra features that are really useful in understanding the prototypes that are hidden in the database. In other words, it adds the optimal amount of complexity. Dropping any information from the MRG would amount to oversimplifying the representation; adding more information would amount to messing it up with irrelevant details.

4.4   Graph In-Out

Another approach to visualize the output obtained from datasets that are transformed by ANNS is what we call a Graph IN-OUT (G IN-OUT). The G IN-OUT reconfigures the output of an ANN (Auto-CM) so that it is able to handle non-symmetric matrices and sparse graphs.

4.4.1   Theory

IN-OUT Graph (G IN-OUT) is an algorithm suitable to filter a squared matrix of weights, with null main diagonal, generated by any other algorithm. G IN-OUT algorithm represents the weight matrix that is an output to an ANN, or in our case, Auto-CM, as a weighted graph with specific features:

a.   The resulting graph is a directed graph when the original matrix is not symmetric;
b.   The resulting graph could be composed of sub-graphs not necessary connected each other's (sparse graph).

The logic of G IN-OUT algorithm is simple: If we consider a square asymmetric matrix of connections, where the columns represent the nodes receiving the weighted connections, and the rows represent the same nodes sending their weighted connections, then:

a.   Each node of the rows is linked only to one other node of the columns of its row, where the weight is the strongest: Consequently, each row node will receive an arc from only one column node. If A is on the rows and B is on the columns then the arc goes from B to A: B → A; that is, for each row node there is only one input (OOI);
b.   Each node of the columns is linked to only one node of the rows of its column, where the weights is the strongest; consequently, each column node will send an arc to only one row node. If A is on the columns and B is on the row then the arc goes from A to B: A → B that is, for each column node there is only one output (OOU);

The OOI arc means which node of each column is the closest to which node of each row, while the OOU arc means to which node of the rows is the closest to which node of each column.

We consider the weighted value, W, between two nodes as the **fuzzy intersection** among them found by a specific algorithm:

$$W(A, B) = \mu_{\bar{B}}(A) \text{ and } W(B, A) = \mu_{\bar{A}}(B) \qquad (4.17)$$

where $W(A, B)$ and $W(B, A)$ can be thought to represent the possibility that A belongs to B and the possibility that B belongs to A respectively. When these two symmetric memberships are different, that is,

$$\mu_{\bar{B}}(A) > \mu_{\bar{A}}(B) \text{ or } \mu_{\bar{B}}(A) < \mu_{\bar{A}}(B), \tag{4.18}$$

this means that the possibility that an element of A belongs also to set B is different that the probability of an element of B belongs to set A.

$$pos(x \in A|x \in B) \neq pos(y \in B|y \in A); \tag{4.19}$$

Consequently

$$pos\,(x \in A|x \in B) < pos\,(y \in B|y \in A) \text{ if } \mu_{\bar{B}}(A) < \mu_{\bar{A}}(B). \tag{4.20}$$

or

$$pos\,(x \in A|x \in B) < pos\,(y \in B|y \in A) \text{ if } \mu_{\bar{B}}(A) < \mu_{\bar{A}}(B). \tag{4.21}$$

4.4.2 Example 1—A Simple Case

Let us present an example. Table 4.3a shows the number of phone calls among four actors. The Tables 4.3b, c show the same data from the probabilistic point of view of the caller and of the called. Let us plot the Input Graph (G_IN) considering the value W of the matrix:

a. Actor A calls more frequently actor D;
b. Actor B calls more frequently actor C;
c. Actor C calls more frequently actor A;
d. Actor D calls more frequently actor C.

Figure 4.4 represents the OOI dynamics: Each node (actor) receives one arc from the actor that it finds closest. We suppose that an actor calls more frequently another actor having more shared features with him. Now if this supposition is right, we can write Eqs. (31) or (34) in this form:

$$pos\,(x \in D|x \in A) < pos\,(y \in A|y \in D)$$

That is, the possibility that a "x" feature of A is present in D is smaller than the possibility that a "y" feature of D is present also in A.

Let us plot now the Output Graph (G_OUT):

a. Actor A is called more frequently from actor C;
b. Actor B is called more frequently from actor C;
c. Actor C is called more frequently from actor D;
d. Actor D is called more frequently from actor A.

Table 4.3 **a** A square matrix with null diagonal and values with asymmetric connections between four actors which call each other. **b** The possibility of each sender to call another actor. **c** The possibility of each sender to be called by another actor

Calls frequencies

Sender	Receiver				
		A	B	C	D	Total
	A	0	1	0	9	10
	B	0	0	5	3	8
	C	8	4	0	3	15
	D	2	3	7	0	12
	Total	10	8	12	15	

Sender probability

Sender	Receiver				
		A	B	C	D	Total
	A	0	0.1	0	0.9	1
	B	0	0	0.625	0.375	1
	C	0.533333	0.266667	0	0.2	1
	D	0.166667	0.25	0.583333	0	1

Receiver probability

Sender	Receiver			
		A	B	C	D
	A	0	0.125	0	0.6
	B	0	0	0.416667	0.2
	C	0.8	0.5	0	0.2
	D	0.2	0.375	0.583333	0
	Total	1	1	1	1

Figure 4.5 represents the OOU dynamics: each node (actor) sends one arc to the actor that finds it closest. Figure 4.6 shows the Graph In and the Graph Out together (G_IN-OUT).

4.4.3 Example 2—A More Complex Case, Natural Clustering

Let us figure out the average number of phone calls among 500 subjects in one week. We have a square asymmetric matrix 500 × 500, with null diagonal. We have generated randomly all these calls, but we have embedded into the random matrix three hidden groups:

a. The first group, named the red's group, is composed of five individuals;
b. The second group, named the green's group, composed of four individuals;
c. The third group, named the yellow's group, is composed of three individuals.

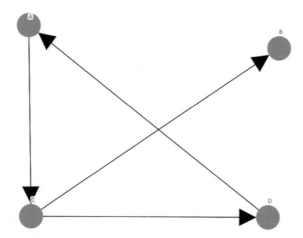

Fig. 4.4 Graph IN of the matrix shown in Table 4.3a

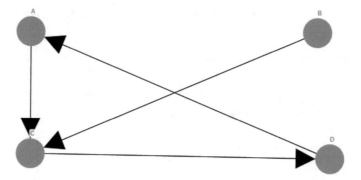

Fig. 4.5 Graph OUT of the matrix shown in Table 4.3b

The individuals into each group communicate among themselves about two times more intensively than in the other communications exchanges. That is, the random probability that each member of a group may call another member of the same group is double. Table 4.2 shows the scores sorted by average frequency of each individual of the three hidden groups with the first 10 actors of the dataset. From Table 4.2 is evident that the data by themselves do not allow clustering and does not allow the discovery of the right group using a simple sorting technique (Table 4.4).

We have applied to the same matrix the G_IN and the G_OUT algorithm and then we have plotted the 500 Actors with the colour of their group (the members of no group are in cyan). Figure 4.7 shows the G_IN and Fig. 4.8 shows the G_OUT.

Table 4.4   The top five closest members of the hidden group

The top five rankings					
3 of 4 in the list Actor_89[][]	Actor_96[][] 13.30	Actor_299[][] 10.72	Actor_424[][] 10.41	Actor_5 9.99	Actor_121 9.96
3 of 4 in the list Actor_96[][]	Actor_89[][] 13.22	Actor_299[][] 10.97	Actor_424[][] 10.45	Actor_394 9.99	Actor_445 9.95
0 of 3 in the list Actor_113**	Actor_108 9.97	Actor_233 9.96	Actor_310 9.89	Actor_63 9.87	Actor_292 9.86
4 of 5 in the list Actor_142<>	Actor_387<> 15.77	Actor_296<> 11.47	Actor_181<> 10.81	Actor_83 10.00	Actor_322<> 9.99
2 of 3 in the list Actor_181<>	Actor_322<> 15.30	Actor_296<> 11.80	Actor_242 9.98	Actor_17 9.97	Actor_264 9.96
2 of 5 in the list Actor_209**	Actor_113** 16.70	Actor_441** 13.41	Actor_153 9.98	Actor_167 9.98	Actor_425 9.93
3 of 5 in the list Actor_296<>	Actor_142<> 14.82	Actor_387<> 11.29	Actor_322<> 11.09	Actor_369 10.00	Actor_102 9.99
1 of 4 in the list Actor_299[][]	Actor_89[][] 16.96	Actor_409 9.98	Actor_294 9.95	Actor_363 9.94	Actor_381 9.94
3 of 5 in the list Actor_322<>	Actor_181<> 16.92	Actor_387<> 14.18	Actor_142<> 11.38	Actor_374 9.99	Actor_276 9.99
3 of 5 in the list Actor_387<>	Actor_322<> 15.35	Actor_181<> 14.14	Actor_296<> 12.19	Actor_178 9.99	Actor_129 9.99
2 of 4 in the list Actor_424[][]	Actor_299[][] 17.94	Actor_89[][] 17.34	Actor_88 9.97	Actor_183 9.97	Actor_85 9.96
1 of 3 in the list Actor_441**	Actor_113** 14.17	Actor_77 10.00	Actor_306 9.99	Actor_392 9.99	Actor_102 9.97

[][] = Green Group
<> = Red Group
** = Yellow Group

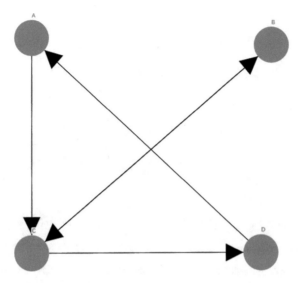

Fig. 4.6 Graph IN-OUT of the matrix shown in Table 4.3c

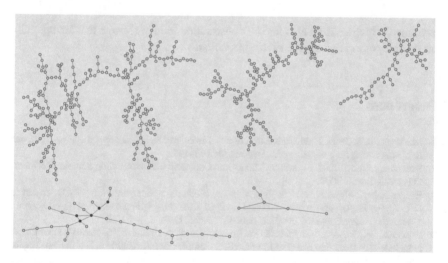

Fig. 4.7 Graph IN of the 500 actors. The members of the hidden groups are enhanced with their typical colours

4.5 Summary

This chapter presented three ways to represent the final Auto-CM weights as a graph. The first, MST works with the final matrix as is. The MRG uses the MST to find by construction complex structures in the MST. The G IN-OUT graphs are used for

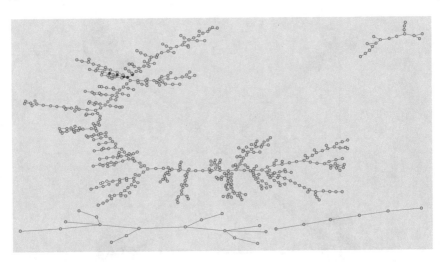

Fig. 4.8 Graph OUT of the 500 actors. The members of the hidden groups are enhanced with their typical colours

asymmetric matrices and also find by construction hidden clusters that are typically impossible to find by usual clustering techniques.

References

1. Mantegna, R.N. 1999. Hierarchical structure in Financial Market. *European Physical Journal B: Condensed Matter and Complex Systems* 11: 193–197.
2. Buscema, M (ed.). 2007. Squashing Theory and Contractive Map Network, Semeion Technical Paper #32, Rome, 2007.
3. Buscema, M. 2007. A Novel Adapting Mapping Method for Emergent Properties Discovery in Data Bases: Experience in Medical Field. In *2007 IEEE International Conference on Systems, Man and Cybernetics (SMC 2007)*. Montreal, Canada, October 7–10, 2007.
4. Buscema, M., M. Asadi-Zeydabadi, W. Lodwick, and M. Breda. 2016. The H0 Function, A New Index for Detecting Structural/Topological Complexity Information in Undirected Graphs. *Physica A* 447: 355–378.
5. McClellan, J.L., and D.E. Rmelhart. 1986. *Explorations in Parallel Distributed Processing*. Cambridge, MA: MIT Press.
6. Buscema, M., V. Consonni, D. Ballobio, A. Mauri, G. Massini, M. Breda, and R. Todeschini. 2014. K-CM: A New Artificial Neural Network. Applications to Supervised Patter Recogniction. *Chemometrics and Intetelligent Systems* 138: 110–119.

Chapter 5
Dataset Transformations and Auto-CM

Abstract We have looked at how to visualize the relationships among the elements of a dataset in Chap. 4. This chapter is devoted to the use of Auto-CM in the transformation of datasets for the purpose of extracting further relationships among data elements. The first transformation we call the FS-Transform, which looks beyond an all or nothing, binary relationship that is typical of most ANNs. The extraction of these perhaps more subtle relationships can be thought of as gradual relationships, zero denoting no relationship is present and one denoting a full/complete relationship that is absolutely present. It is thus, akin to a fuzzy set. The second transformation is one, which "morph" the delineation between records and variables within records that we call Hyper-Composition.

5.1 Introduction

The output from Auto-CM, at the end of training, can be used to rewrite a modified version whole dataset. This new dataset, in many cases, may present a more useful set. In fact, we can take the record and variable values in a dataset and use the output of Auto-CM to transform the dataset, taking into account of the distribution of the all other values in every other variable and record. We consider this transformed dataset as one where the variables may be considered to relate to one another in degrees or relating on a scale of zero to one. Typically the state of a system is either activated or it is not, that is, on or off. When a state is partially activated between zero (inactive or off) and one (active or on), it is considered in our approach to be fuzzy which is consistent with the way this term is used. A state may be in transition between active and inactive. A dataset that is transformed into relationships described by such transitions becomes one in which there are fuzzy relationships. In addition, we show how to use the output from Auto-CM to rewrite the database that considers records and variables together.

© Springer International Publishing AG, part of Springer Nature 2018
P. M. Buscema et al., *Artificial Adaptive Systems Using Auto Contractive Maps*, Studies in Systems, Decision and Control 131, https://doi.org/10.1007/978-3-319-75049-1_5

Fig. 5.1 The new topology
of FS-transform
auto-contractive map

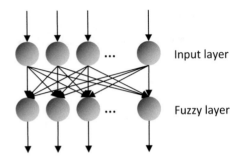

5.2   FS-Transform Auto-CM

The method by which the output from Auto-CM is transformed by taking into consideration the influence of all other variables in all records of the original database we call ***FS-Transform Auto-CM*** (see [1] where, here, we have changed nomenclature from Profiling to FS-transform). The FS-transform uses a different topology than that of the Auto-CM topology. The FS-transform uses only two layers completely connected through the trained weights matrix. The hidden layer of the regular Auto-C M ANN will be the FS-transform whereas Auto-CM is input layer, and the original output layer is transformed into a fuzzy layer (see Fig. 5.1).

The transformation from the input layer to the fuzzy layer is regulated according to Eq. (5.1).

$$z_i^{[p]} = \mu_i\left(x_p\right) = x_i^{[p]}\left(1 - \frac{\sum_{j=1}^{N} x_j^{[p]}\left(1 - w_{i,j}^{[T]}\right)}{C^2}\right) \tag{5.1}$$

where:

$x_i^{[p]}$ The i-th variable of the p-th pattern of the original dataset;

$w_{i,j}^{[T]}$ The trained Hidden-Output weights matrix of Auto-CM;

C The usual Constant of Auto-CM used for the training phase;

$z_i^{[p]}$ The fuzzified new output for the i-th variable of the p-th record of the original dataset;

$\mu_i\left(x_p\right)$ The membership function for the fuzzy set $z_i^{[p]}$.

In the above, [p] means the pth pattern and T means training. Equation (5.1) we call the *FS-transform* and it essentially fuzzifies the input values of the original dataset by means the weight of the full grid of connections as represented by the matrix $W = (w_{i,j})$ generated by the regular Auto-CM ANN at the end of the training phase. The new output value, $z_i^{[p]}$ indicates the degree to which the pth record belongs to the ith variable, that is its fuzzy membership value that is given in Eq. (5.1).

This transformation generates a new dataset where the original values of the training dataset are reformulated in a fuzzy way. This new fuzzy dataset, a different

Table 5.1 Dataset of Jets and Sharks

Name	Gang	Age	Education	Status	Occupations
ART	Jets	40	Junior School	Single	Pusher
AL	Jets	30	Junior School	Married	Burglar
SAM	Jets	20	College	Single	Bookie
CLYDE	Jets	40	Junior School	Single	Bookie
MIKE	Jets	30	Junior School	Single	Bookie
JIM	Jets	20	Junior School	Divorced	Burglar
GREG	Jets	20	High School	Married	Pusher
JOHN	Jets	20	Junior School	Married	Burglar
DOUG	Jets	30	High School	Single	Bookie
LANCE	Jets	20	Junior School	Married	Burglar
GEORGE	Jets	20	Junior School	Divorced	Burglar
PETE	Jets	20	High School	Single	Bookie
FRED	Jets	20	High School	Single	Pusher
GENE	Jets	20	College	Single	Pusher
RALPH	Jets	30	Junior School	Single	Pusher
PHIL	Sharks	30	College	Married	Pusher
IKE	Sharks	30	Junior School	Single	Bookie
NICK	Sharks	30	High School	Single	Pusher
DON	Sharks	30	College	Married	Burglar
NED	Sharks	30	College	Married	Bookie
KARL	Sharks	40	High School	Married	Bookie
KEN	Sharks	20	High School	Single	Burglar
EARL	Sharks	40	High School	Married	Burglar
RICK	Sharks	30	High School	Divorced	Burglar
OL	Sharks	30	College	Married	Pusher
NEAL	Sharks	30	High School	Single	Bookie
DAVE	Sharks	30	High School	Divorced	Pusher

view of the information contained in the original dataset, is used to train the regular Auto-CM ANN.

Next we discuss what Eq. (5.1) does to the data and why it gives us a useful view of the dataset. Let us examine the effect of the FS-transform equations of Auto-CM by taking our example of the dataset called "Jet and Sharks" gangs (henceforth referred to as "Gang 14×27"), seen in Chap. 4. We encode the five variables into 14 binary variables (2 gangs + 3 groups of ages + 3 levels of education + 3 marital status designations + 3 occupations). Table 5.1 shows the original database and Table 5.2 shows the statistical distribution of each variable in each of the two gangs (Jet and Sharks). Table 5.3 shows the new binary dataset.

Table 5.2 Statistics of the
Jets and Sharks dataset

Variable	Jet	Shark	Jet (%)	Shark (%)
20s	9	1	60.00	8.33
30s	4	9	26.67	75.00
40s	2	2	13.33	16.67
JH	9	1	60.00	8.33
HS	4	7	26.67	58.33
COL	2	4	13.33	33.33
Single	9	4	60.00	33.33
Married	4	6	26.67	50.00
Divorced	2	2	13.33	16.67
Pusher	5	4	33.33	33.33
Bookie	5	4	33.33	33.33
Burglar	5	4	33.33	33.33

Table 5.4 shows the dataset "rewritten" by the FS-Transform equations of Auto-CM. The membership of each variable and each record was reduced to a value between 0 and 1, as in fuzzy set theory. The new membership values, for each variable-class, implies that the entire dataset is not considered a simple "list" of records, but it constitutes a system in which each new value (the degree of membership) involves an analysis of the distribution in terms of sensitivity and specificity with respect to all other values.

Remarks: The record "IKE" is classified in the original dataset with the title "Junior School." However, "IKE" is the only member of the Sharks to present this feature, monopolized by members of the Jets gang (ratio 1/9). In fact, previous studies show this record to be an outlier [2]. FS-Transform applied in conjunction with Auto-CM, consequently, assigns to "IKE" the lowest value of membership to the class "Junior School" and also a low value to belonging to the "Sharks". The same characterization of FS-Transform run in conjunction with Auto-CM reserves for the record "KEN", defined in previous studies as an outlier [2]. This is the only subject of the Sharks defined as "20's". Again his belonging to this class has the lowest value and also its representation in the Sharks is the lowest ever.

The effectiveness of the equations FS-Transform stands out from the comparison of the minimal spanning tree constructed from the Euclidean distance of the original dataset (see Fig. 5.2) with the one built by the Euclidean distance of the dataset reconstructed from FS-Transform equations (see Fig. 5.3). The position of the two outlier records are in a dubious position as shown in Fig. 5.2 while the FS-Transform groups correctly all of the records according to their membership in each of the two gangs (see again Fig. 5.3).

We could make many other observations related to the transformed dataset but suffice it to say that the FS-Transform equations transform a simple list of records into a system of fuzzy memberships as given by Eq. (5.2).

Table 5.3 Pre-processing of Jets and Sharks dataset

Gang 14×27	Jet	Sharks	20's	30's	40's	JH	COL	HS	Single	Married	Divorced	Pusher	Bookie	Burglar
ART	1	0	0	0	1	1	0	0	1	0	0	1	0	0
AL	1	0	0	1	0	1	0	0	0	1	0	0	0	1
SAM	1	0	1	0	0	0	1	0	1	0	0	0	1	0
CLYDE	1	0	0	0	1	1	0	0	1	0	0	0	1	0
MIKE	1	0	0	1	0	1	0	0	1	0	0	0	1	0
JIM	1	0	1	0	0	1	0	0	0	0	0	0	0	0
GREG	1	0	1	0	0	0	0	1	0	1	0	1	0	1
JOHN	1	0	1	0	0	1	0	0	0	1	0	0	0	0
DOUG	1	0	0	1	0	0	0	1	1	0	0	0	1	1
LANCE	1	0	1	0	0	0	0	0	0	1	1	0	0	0
GEORGE	1	0	1	0	0	1	0	0	0	0	0	0	0	1
PETE	1	0	1	0	0	0	0	1	1	0	0	0	1	1
FRED	1	0	1	0	0	0	0	1	1	0	0	1	0	0
GENE	1	0	1	0	0	0	1	0	1	0	0	1	0	0
RALPH	1	0	0	1	0	1	0	0	1	1	0	1	0	0
PHIL	0	1	0	1	0	0	1	0	0	0	0	1	1	0
IKE	0	1	0	1	0	1	0	0	1	0	0	0	0	0
NICK	0	1	0	1	0	0	0	1	1	1	0	1	0	0
DON	0	1	0	1	0	0	1	0	0	1	0	0	1	1
NED	0	1	0	1	0	0	1	0	0	1	0	0	1	0
KARL	0	1	0	0	1	0	0	1	0	0	0	0	0	0
KEN	0	1	1	0	0	0	0	1	1	1	0	0	0	1
EARL	0	1	0	0	1	0	0	1	0	0	0	0	0	1
RICK	0	1	0	1	0	0	0	1	0	1	1	0	0	1
OL	0	1	0	1	0	0	1	0	0	0	0	1	0	0
NEAL	0	1	0	1	0	0	0	1	1	0	0	0	1	0
DAVE	0	1	0	1	0	0	0	1	0	0	1	1	0	0

Table 5.4 The Gang fuzzified dataset

Gang dataset	Jet	Sharks	20's	30's	40's	JH	COL	HS	Single	Married	Divorced	Pusher	Bookie	Burglar
ART	0.8758	0.0000	0.0000	0.0000	0.7992	0.8592	0.0000	0.0000	0.8681	0.0000	0.0000	0.8370	0.0000	0.0000
AL	0.8695	0.0000	0.0000	0.8552	0.0000	0.8813	0.0000	0.0000	0.0000	0.8709	0.0000	0.0000	0.0000	0.8750
SAM	0.8758	0.0000	0.8596	0.0000	0.0000	0.0000	0.8119	0.0000	0.8766	0.0000	0.0000	0.0000	0.8544	0.0000
CLYDE	0.8757	0.0000	0.0000	0.0000	0.8125	0.8684	0.0000	0.0000	0.8758	0.0000	0.0000	0.0000	0.8623	0.0000
MIKE	0.8894	0.0000	0.0000	0.8686	0.0000	0.8807	0.0000	0.0000	0.8986	0.0000	0.0000	0.0000	0.8781	0.0000
JIM	0.8763	0.0000	0.8772	0.0000	0.0000	0.8766	0.0000	0.0000	0.0000	0.0000	0.0000	0.0000	0.0000	0.8773
GREG	0.8698	0.0000	0.8710	0.0000	0.0000	0.0000	0.0000	0.8583	0.0000	0.8501	0.8229	0.8607	0.0000	0.0000
JOHN	0.8904	0.0000	0.0848	0.0000	0.0000	0.8836	0.0000	0.0000	0.0000	0.8618	0.0000	0.0000	0.0000	0.8890
DOUG	0.8689	0.0000	0.0000	0.8735	0.0000	0.0000	0.0000	0.8740	0.9032	0.0000	0.0000	0.0000	0.8852	0.0000
LANCE	0.8904	0.0000	0.8848	0.0000	0.0000	0.8836	0.0000	0.0000	0.0000	0.8618	0.0000	0.0000	0.0000	0.8890
GEORGE	0.8763	0.0000	0.8772	0.0000	0.0000	0.8766	0.0000	0.0000	0.9008	0.0000	0.8229	0.0000	0.8667	0.8773
PETE	0.8898	0.0000	0.8738	0.0000	0.0000	0.0000	0.0000	0.8707	0.9008	0.0000	0.0000	0.8703	0.0000	0.0000
FRED	0.8899	0.0000	0.8823	0.0000	0.0000	0.0000	0.0000	0.8710	0.8932	0.0000	0.0000	0.8655	0.0000	0.0000
GENE	0.8758	0.0000	0.8681	0.0000	0.0000	0.0000	0.8204	0.0000	0.8689	0.0000	0.0000	0.8662	0.0000	0.0000
RALPH	0.8895	0.0000	0.0000	0.8609	0.0000	0.8715	0.0000	0.0000	0.8910	0.0000	0.0000	0.8644	0.0000	0.0000
PHIL	0.0000	0.0890	0.0000	0.8869	0.0000	0.0000	0.8667	0.0000	0.0000	0.8785	0.0000	0.0000	0.0000	0.0000
IKE	0.0000	0.8539	0.0000	0.8898	0.0000	0.8381	0.0000	0.0000	0.8813	0.0000	0.0000	0.0000	0.8766	0.0000
NICK	0.0000	0.8925	0.0000	0.8952	0.0000	0.0000	0.0000	0.8916	0.8782	0.0000	0.0000	0.8787	0.0000	0.0000
DON	0.0000	0.8907	0.0000	0.8760	0.0000	0.0000	0.8453	0.0000	0.0000	0.8913	0.0000	0.0000	0.8467	0.8455
NED	0.0000	0.8900	0.0000	0.8865	0.0000	0.0000	0.8582	0.0000	0.0000	0.8706	0.0000	0.0000	0.8432	0.0000
KARL	0.0000	0.8720	0.0000	0.0000	0.8200	0.0000	0.0000	0.8617	0.0000	0.8479	0.0000	0.0000	0.0000	0.8452
KEN	0.0000	0.8507	0.8526	0.0000	0.0000	0.0000	0.0000	0.8824	0.8467	0.0000	0.0000	0.0000	0.0000	0.8452
EARL	0.0000	0.8727	0.0000	0.0000	0.8074	0.0000	0.0000	0.8563	0.0000	0.8685	0.0000	0.0000	0.0000	0.8487
RICK	0.0000	0.8799	0.0000	0.8613	0.0000	0.0000	0.0000	0.8634	0.0000	0.0000	0.8238	0.0000	0.0000	0.8536
OL	0.0000	0.8890	0.0000	0.8869	0.0000	0.0000	0.8667	0.0000	0.0000	0.8785	0.0000	0.8644	0.0000	0.0000
NEAL	0.0000	0.8935	0.0000	0.8948	0.0000	0.0000	0.0000	0.8912	0.8858	0.0000	0.0000	0.0000	0.8838	0.0000
DAVE	0.0000	0.8783	0.0000	0.8722	0.0000	0.0000	0.0000	0.8692	0.0000	0.0000	0.8013	0.8496	0.0000	0.0000

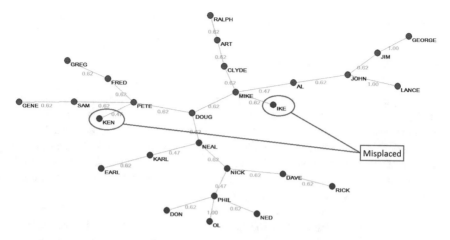

Fig. 5.2 MST based on the Euclidean distance of the original dataset

Fig. 5.3 MST based on the Euclidean distance of the fuzzified data using FS-transform

$$R_i \left(v_{i,1}, v_{i,2}, \ldots, v_{i,N} \right) \Rightarrow R_i^{[F]} \left(\mu_{i,1} \left(x_1 \right), \mu_{i,2} \left(x_2 \right), \ldots, \mu_{i,N} \left(x_N \right) \right), \qquad (5.2)$$

where

$R_i \left(v_{i,1} v_{i,2}, \ldots, v_{i,N} \right)$ original record defined by N variables;

$R_i^{[F]} \left(\mu_{i,1} \left(x_1 \right), \mu_{i,2} \left(x_2 \right), \ldots, \mu_{i,N} \left(x_N \right) \right)$ new fuzzified record defined by N new memberships.

5.3   Hyper-Composition—Views of Variables and Records

5.3.1   Introduction to Hyper-Composition

It is often useful to display a dataset that is made up of records (rows) where the rows are composed of attributes, variable elements (columns) as a whole with records and attributes together in one analysis. We use the Maximal Regular Graph as a display. Our context is one in which we first find the relationships underlying a dataset, one display of which is the associated MRG that depicts these relationships as a graph of the dataset relationships with connections between the nodes that represent these relationships. This is the "graph" part of MRG.

The regular part is the degree of the vertices (nodes which will be the attributes/variables). Recall that the degree of a vertex is its number of incident edges. A maximal graph is a graph with a given property (its degree in our case) for which a given set of vertices have with respect to their connections (edges) such that it is not possible to add one more edge for that given set of vertices. In other words, the MRG is a graph for a given set of vertices for which the degree is maximal. So, given a set of vertices, it is the rearrangement of the vertices in such a way that the degree of the nodes/vertices is maximal, that is, we have the most connection. This, in our context, means that we have the highest possible connectivity given the configuration of the vertices.

Beyond a view of the relationships that any graph provides, the MRG displays sub-graphs (some completely connected sub-graphs for example) that allow for a "visual clustering". Thus, there are two steps. The first is to transform a dataset composed of records and attributes to one that is the associated record/attribute matrix. The second step is to produce a display of the relationships with a MRG.

The way a graph of a dataset is constructed, via what we call the *Generalized Hyper-Composition algorithm* (GHC), is the subject of this section of the presentation. GHC synthesizes in one matrix, variables and records (coordinates and points).

Let us image a matrix M, where we have Rows(P) = {A, B, C, D} and the Columns(N) = {α, β, γ}. Usually a complete multivariate analysis of the matrix M is possible, using Principal Component Analysis (PCA) or an Auto-Associative Neural Network. In both cases we have to choose if we want to map the N variables, using the records as coordinates in a P dimensional space, or if we want to transpose the matrix M^T, in order to map the P records using the variables as coordinates in the a N dimensional space.

The mapping of *both* simultaneously is very hard, because it is not possible to measure the similarity of variables and records among them using a shared metric. It is analogous to trying to put on a glove at the same moment at which this glove is being worn by the same hand on which it is being put. It seems to be a problem of circularity.

The bi-plot of PCA seems to resolve this problem using a linear hyper-plane optimizing the global variance and rearranging the original variables in new ranked components. Generalized Hyper Composition Auto-CM is a procedure able to provide a positive answer to this complex problem in a nonlinear rather than a linear way.

5.3.2 The Auto-CM Neural Network—Review

Auto-CM is used as basic algorithm to build up the hyper-composition matrix. Auto-CM was already presented in many theoretical in Chap. 3 (application of Auto-CM can be found in [1, 3–11]). Recall the basic equation of Auto-CM learning process from Chap. 3 which we repeat below [see Eqs. (5.3)–(5.10)].

$$m_{i,p(n)}^{[h]} = m_{i,p(n)}^{[s]} \cdot \left(1 - \frac{v_{i(n)}}{C}\right) \tag{5.3}$$

$$\Delta v_{i(n)} = \sum_{p}^{M} \left(m_{i,p(n)}^{[s]} - m_{i,p(n)}^{[h]}\right) \cdot \left(1 - \frac{v_{i(n)}}{C}\right) \cdot m_{i,p(n)}^{[s]} \tag{5.4}$$

$$v_{i(n+1)} = v_{i(n)} + \alpha \cdot \Delta v_{i(n)} \tag{5.5}$$

$$Net_{i,p(n)} = \frac{1}{N} \sum_{j=1}^{N} \left(m_{i,p(n)}^{[h]}\right) \cdot \left(1 - \frac{w_{i,j(n)}}{C}\right); \tag{5.6}$$

$$m_{i,p(n)}^{[t]} = m_{i,p(n)}^{[h]} \cdot \left(1 - \frac{Net_{i,p(n)}}{C^2}\right); \tag{5.7}$$

$$\Delta w_{i,j(n)} = \sum_{p}^{M} \left(m_{i,p(n)}^{[h]} - m_{i,p(n)}^{[t]}\right) \cdot \left(1 - \frac{w_{i,j(n)}}{C}\right) \cdot m_{j,p(n)}^{[h]} \tag{5.8}$$

$$w_{i,j(n+1)} = w_{i,j(n)} + \alpha \cdot \Delta w_{i,j(n)}. \tag{5.9}$$

The convergence condition is:

$$\lim_{n \to \infty} \Delta w_{i,j(n)} = 0, \forall v_{i(n)} = C \tag{5.10}$$

For Eqs. (5.3)–(5.9), $m_{i,p(n)}^{[s]}$, $m_{i,p(n)}^{[h]}$ and $m_{i,p(n)}^{[t]}$ are input, hidden and output units respectively, $v_{i(n+1)}$ and $w_{i,j(n)}$ are input-hidden and hidden-output weight respectively, N and M are number of inputs and patterns, C is a constant and it is typically $C = \sqrt{N}$, and $i, j \in [1, 2, \ldots N]$, $p \in [1, 2, \ldots M]$, n is number of epochs and α is the learning coefficient.

At the end of the learning phase, the fundamental knowledge of the given dataset is embedded in the trained weights matrix of the second layer, **W**, of the Auto-CM.

5.3.3 The GHC Algorithm

Let us consider a dataset with N variables and P records (a N × P matrix). Auto-CM can be applied to the dataset to find the global nonlinear association among the variables (many-to-many relations). The weights trained matrix, $\mathbf{W}$, synthesize the information that Auto-CM is able to extract from the dataset.

We now multiply the weights matrix $\mathbf{W}$ with a "quasi identity" matrix in the following way:

Auto CM Weights ($\mathbf{W}$)   Quasi-Identity Matrix ($\mathbf{I}$)

$$
\begin{pmatrix}
w_{1,1} & w_{1,2} & w_{1,3} & w_{1,\ldots} & w_{1,N} \\
w_{2,1} & w_{2,2} & w_{2,3} & w_{2,\ldots} & w_{2,N} \\
w_{3,1} & w_{3,2} & w_{3,3} & w_{3,4} & w_{3,N} \\
w_{\ldots,1} & w_{\ldots,2} & w_{\ldots,3} & w_{\ldots,\ldots} & w_{\ldots,N} \\
w_{N,1} & w_{N,2} & w_{N,3} & w_{N,\ldots} & w_{N,N}
\end{pmatrix}
\times
\begin{pmatrix}
1 & 10^{-6} & 10^{-6} & 10^{-6} & 10^{-6} \\
10^{-6} & 1 & 10^{-6} & 10^{-6} & 10^{-6} \\
10^{-6} & 10^{-6} & 1 & 10^{-6} & 10^{-6} \\
10^{-6} & 10^{-6} & 10^{-6} & 1 & 10^{-6} \\
10^{-6} & 10^{-6} & 10^{-6} & 10^{-6} & 1
\end{pmatrix}
$$

Results ($\mathbf{Out}$)

$$
=
\begin{pmatrix}
Out_{1,1} & Out_{1,2} & Out_{1,3} & Out_{1,\ldots} & Out_{1,N} \\
Out_{2,1} & Out_{2,2} & Out_{2,3} & Out_{2,\ldots} & Out_{2,N} \\
Out_{3,1} & Out_{3,2} & Out_{3,3} & Out_{3,4} & Out_{3,N} \\
Out_{\ldots,1} & Out_{\ldots,2} & Out_{\ldots,3} & Out_{\ldots,\ldots} & Out_{\ldots,N} \\
Out_{N,1} & Out_{N,2} & Out_{N,3} & Out_{N,\ldots} & Out_{N,N}
\end{pmatrix}
$$

In other words, the output matrix ($\mathbf{Out}$) is generated by the following equation (see Eq. (5.11) as the output of this process):

$$
Out_{i,j} = l\left(\sum_{k}^{N} I_{i,j,k} \cdot w_{i,k}\right)
$$

That is,

$$Out_{1,1} = l\left(I_{1,1}w_{1,1} + I_{1,2}w_{1,2} + \cdots + I_{1,N}\right)$$
$$Out_{1,2} = l\left(I_{1,1}w_{2,1} + I_{1,2}w_{2,2} + \cdots + I_{1,N}\right)$$
$$\cdots$$
$$Out_{1,N} = l\left(I_{1,1}w_{N,1} + I_{1,2}w_{N,2} + \cdots + I_{1,N}\right)$$
$$\cdots$$
$$\cdots$$
$$Out_{2,1} = l\left(I_{2,1}w_{1,1} + I_{2,2}w_{1,2} + \cdots + I_{2,N}\right)$$
$$Out_{2,2} = l\left(I_{2,1}w_{2,1} + I_{2,2}w_{2,2} + \cdots + I_{2,N}\right)$$
$$\cdots$$
$$Out_{2,N} = l\left(I_{2,1}w_{N,1} + I_{2,2}w_{N,2} + \cdots + I_{2,N}\right) \tag{5.11}$$

So, in order to attain $Out_{i,j}$, we fix the ith row in the Quasi-Identity matrix, then we consider the inner product between itself and the jth row in the **W** matrix. At the end a function of linear scaling is applied on the result.

The following pseudo-code explains how it works:

```
for(i=0; i<NumInput; i++)
{
        for(q=0; q<NumInput; q++)
        {
                if(q==i) Input[q]=1; else Input[q]=0;
        }

        for(j=0; j<NumInput; j++)
        {
                Net=0;
                for(k=0; k<NumInput; k++)
                {
                        Net+=Input[k]*w[j][k];
                }
                Out[i][j]=Linear_Scaling(Net);
        }
}
```

Now, if we want to understand how the records are activated when a combination of variables are inputted respond to any call of each variable, we need transform each output of Eq. (5.11), from which the variables are activated from the other variables, into a weight in the interval $[-1, +1]$ and then we need to weight this value with the original value that each single record presents to all the variables of the original dataset [see Eqs. (5.12)–(5.13)]. This is done to understand how each variable is activated from the others. Subsequently upon knowing the activation values, for each variable, this value is used as a weight to define the activation value of each record of the dataset in relation with all the variables.

$$Net_{k,z} = \sum_{1}^{N} \left(2 \cdot Out_{z,j} - 1\right) R_{k,j} \tag{5.12}$$

$$Final\_Out_{k,z} = \frac{e^{Net_{k,z}} - e^{-Net_{k,z}}}{e^{Net_{k,z}} + e^{-Net_{k,z}}} \tag{5.13}$$

where

$R_{k,j}$ Value of the k-th record in the j-th variable of the source dataset;
P Total number of records;
$k \in [1, 2, \ldots P]$ Index for each record of the assigned source dataset;
z, i Indices for each of the N variable;
$Final\_Out_{z,j}$ Output vector of the z-th record in each i-th variable after the recall process

We have used the hyperbolic tangent to bound the interval of the units activation and to obtain a smooth approximation allowing for the derivative to be computationally stable when we calculate changes (derivatives) in the variable values. Moreover, from a probabilistic point of view, the most that changes in the variable values may take is around the central part of a bell curve and a few in proximity of the tails. This is because the first derivative (the changes of values) of the hyperbolic tangent is a bell curve. After Eqs. (5.11)–(5.13) a new matrix with N + P columns (variables and records) and N rows (Variables) is generated.

This new dataset may be processed again with Auto-CM in order to produce a new square and asymmetric weights matrix of associations among variables and records in the same framework resulting in an (N + P) × (N + P) matrix. At this point any graphical representation of this dataset such as the MRG may be used to cluster the records-variables into their underlying associations.

Example 1
Let us look at an example to illustrate the process. Consider the dataset where we are going to analyze three bits of 0 and 1 as depicted in Table 5.5.

The records are patterns of bits Pat_xxx, and the attributes or variables are the Bit_y, the bit positions within the patterns. Figure 5.4 is a synthetic summary of the GHC process.

Table 5.5 Bit pattern

Dataset	Bit_1	Bit_2	Bit_3
Pat_000	0	0	0
Pat_001	0	0	1
Pat_010	0	1	0
Pat_011	0	1	1
Pat_100	1	0	0
Pat_101	1	0	1
Pat_110	1	1	0
Pat_111	1	1	1

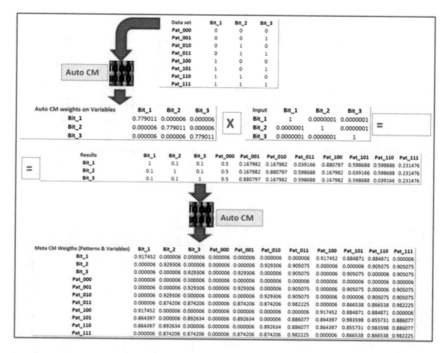

Fig. 5.4   Flow chart of GHC processing the 3 bit data set

The last matrix of Fig. 5.4 (the resulting Auto-CM weights) shows the merged association between variables, records and records with variables. If we apply to this matrix the MRG, we can visualize records and variables in the same weighted graph (see Fig. 5.5).

The red dots are the attributes/variables and the green dots are the records.

1. The records and variables are displayed together.
2. The edges that come out of the nodes representing the variables are precisely the records that have the corresponding bits on.

The only record that has no bit on is Pat_000 which is linked to the variable Bit_1 which has a weight 0, that is, it is not linked and this record is isolated.

Example 2: Rooms dataset
"Rooms" dataset (see [12] page 63) is an artificial dataset composed by the frequency of 50 house objects clustered into 6 types of rooms (see Table 5.6).

We have used GHC algorithm to project this dataset into a weighted graph where the 50 objects and the 6 types of rooms are linked together. We have filtered the weights found by GHC algorithm with the MRG (see Fig. 5.6).

Table 5.6 The rooms dataset

Rooms (50×6)	Restroom	Kitchen	Bedroom	Dinning room	Office	Hall
Desk	0	0	0	0	1	0
Ashtray	0.444	0.222	0.667	0.889	0.889	0.667
Chairs	0	0.889	0.222	0.889	1	0
Easy-chair	0.222	0.889	0.222	0	0.111	0.444
Books	0	0.444	0.444	0.889	0.778	0.778
Bookshelf	0	0.444	0.111	0.889	1	0.778
Pens	0	0.556	0.222	0.444	1	0.444
Papers	0	0.556	0.111	0.444	1	0.444
Soap	1	1	0	0	0	0
Phone	0.111	0.333	0.889	0.889	1	0.667
TV	0	0.444	0.778	0.889	0.556	0
Radio	0.556	0.556	0.778	0.333	0.556	0
Hi-fi	0	0	0	0.778	0.222	0
Computer	0	0	0	0.111	0.889	0
Disks	0	0	0	0.111	0.889	0
Floor-lamp	0	0.222	1	0.222	1	0
Pictures	0	0.222	0.667	1	0.556	1
Clock	0.667	0.889	1	1	1	0.778
Sofa	0	0	0	1	0.111	0
Windows	0.556	1	1	1	1	0.333
Basket	1	1	0	0	1	0.444
Gadgets	0.556	0.444	0.667	0.778	0.333	0.556
Ceiling	1	1	1	1	1	1
Drapes	0.222	0.778	1	0.889	0.667	0.444
Stove	0	1	0	0	0	0.667
Toilet	1	0	0	0	0	0
Walls	1	1	1	1	1	1
Medium	0.444	0.889	0.889	0.333	0.667	0.667
Bed	0	0	1	0	0	0
Arm-chair	0	0	0	0.889	1	0
Shower	1	0	0	0	0	0
Sink	1	1	0	0	0	0
Scale	0.778	0.889	0	0	0	0
Door	1	0.667	1	0.444	1	0.444
Small	0.889	0.667	0.111	0	0.111	0.444
Type-writer	0	0	0	0	1	0
Coffee-cup	0	1	0	0	0	0
coffee pot	0	1	0	0	0	0

(continued)

Table 5.6 (continued)

Rooms (50×6)	Restroom	Kitchen	Bedroom	Dinning room	Office	Hall
Dresser	0.222	0	1	0	0	0.889
Oven	0	0.889	0	0	0	0
Very-small	0.778	0.444	0	0	0.111	0.222
Bookcase	0	0	0.111	0.778	0.889	0.778
Refrigerator	0	1	0	0	0	0
Very-large	0.222	0.111	0.444	0.889	0.889	0.222
Carpet	0	0	0.889	0.889	0.556	0.667
Fireplace	0	0	0	0.778	0	0
Toaster	0	0.889	0	0	0	0
Bathtub	0.778	0	0	0	0	0
Clothes-hanger	1	0	0	0.444	0.667	1
Drawers	0.778	1	1	0.444	1	0.667

Figure 5.6 shows many key bits of information of this dataset:

a. Section 1 clusters the restroom with its specific objects;
b. Section 2 describes the objects shared between kitchen and restroom;
c. Section 3 clusters the kitchen with its specific object;

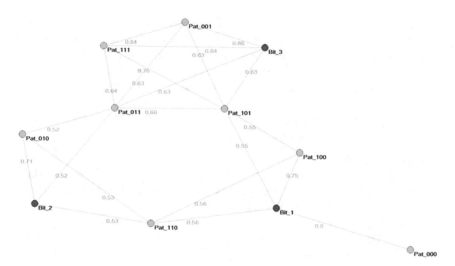

Fig. 5.5 MRG of the 3 bit datasets. Variables (red) and records (green) associated with the same metric

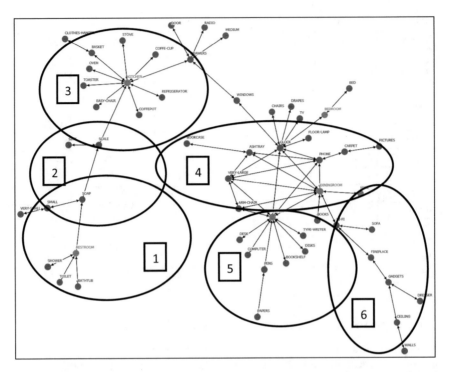

Fig. 5.6   The MRG of the Auto-CM GHC matrix of weights (Red = objects, Green = type of room)

d. Section 4 creates a clique with the objects shared between dining room and the office; it is also interesting how the hall is directly linked to the dining room and outside of the clique the bedroom is directly connected to the "bed";
e. Section 5 clusters only the objects that are typical of the office;
f. Section 6 clusters only the objects that are typical of the dining room;

We may say that the GHC algorithm (and Auto-CM in this case) builds up a very good abstract map of the dataset.

Example 3: The Gang dataset

Gang dataset is a well-known dataset (see [12] page 67), inspired to West Side Story, composed of 27 individuals belonging to two gangs, Jets and Sharks (see Table 5.7). We showed it for the first time in Chap. 4.

The Boolean dataset (Table 5.8) was pre-processed and contains 14 attributes and 27 records. The elementary statistics of the dataset is shown in Table 5.9. We applied the GHC algorithm whose weights are inputted into the MRG (see Fig. 5.7).

Table 5.7 The Gang dataset

Name	Gang	Age	Education	Status	Occupations
ART	Jets	40	Junior School	Single	Pusher
AL	Jets	30	Junior School	Married	Burglar
SAM	Jets	20	College	Single	Bookie
CLYDE	Jets	40	Junior School	Single	Bookie
MIKE	Jets	30	Junior School	Single	Bookie
JIM	Jets	20	Junior School	Divorced	Burglar
GREG	Jets	20	High School	Married	Pusher
JOHN	Jets	20	Junior School	Married	Burglar
DOUG	Jets	30	High School	Single	Bookie
LANCE	Jets	20	Junior School	Married	Burglar
GEORGE	Jets	20	Junior School	Divorced	Burglar
PETE	Jets	20	High School	Single	Bookie
FRED	Jets	20	High School	Single	Pusher
GENE	Jets	20	College	Single	Pusher
RALPH	Jets	30	Junior School	Single	Pusher
PHIL	Sharks	30	College	Married	Pusher
IKE	Sharks	30	Junior School	Single	Bookie
NICK	Sharks	30	High School	Single	Pusher
DON	Sharks	30	College	Married	Burglar
NED	Sharks	30	College	Married	Bookie
KARL	Sharks	40	High School	Married	Bookie
KEN	Sharks	20	High School	Single	Burglar
EARL	Sharks	40	High School	Married	Burglar
RICK	Sharks	30	High School	Divorced	Burglar
OL	Sharks	30	College	Married	Pusher
NEAL	Sharks	30	High School	Single	Bookie
DAVE	Sharks	30	High School	Divorced	Pusher

Figure 5.7 shows a graph with many interesting points:

a. All the members of the two gangs are perfectly separated in the graph (see the black line);
b. The two members connecting the two gangs present the same features but the gang they belong (Neal is a Sharks, Doug is a Jet, see Section 1);
c. The prototypical attributes of the Sharks member is represented in a specific clique (see Section 2);
d. And the members that represent more typically the Sharks gang are clustered in Section 3;

Table 5.8 The transformation of the Gang dataset in Boolean values

Gang 14×7	Jet	Sharks	20's	30's	40's	JH	COL	HS	Single	Married	Divorced	Pusher	Bookie	Burglar
ART	1	0	0	0	1	1	0	0	1	0	0	1	0	0
AL	1	0	0	1	0	1	0	0	0	1	0	0	0	1
SAM	1	0	1	0	0	0	1	0	1	0	0	0	1	0
CLYDE	1	0	0	0	1	1	0	0	1	0	0	0	1	0
MIKE	1	0	0	1	0	1	0	0	1	0	0	0	1	0
JIM	1	0	1	0	0	1	0	0	0	0	1	0	0	1
GREG	1	0	1	0	0	0	0	1	0	1	0	1	0	0
JOHN	1	0	1	0	0	1	0	0	0	1	0	0	0	1
DOUG	1	0	0	1	0	0	0	1	1	0	0	0	1	0
LANCE	1	0	1	0	0	1	0	0	0	1	0	0	0	1
GEORGE	1	0	1	0	0	1	0	0	0	0	1	0	0	1
PETE	1	0	1	0	0	0	0	1	1	0	0	0	1	0
FRED	1	0	1	0	0	0	0	1	1	0	0	1	0	0
GENE	1	0	1	0	0	0	1	0	1	0	0	1	0	0
RALPH	1	0	0	1	0	1	0	0	1	0	0	1	0	0
PHIL	0	1	0	1	0	0	1	0	0	1	0	1	0	0
IKE	0	1	0	1	0	1	0	0	1	0	0	0	1	0
NICK	0	1	0	1	0	0	0	1	1	0	0	1	0	0
DON	0	1	0	1	0	0	1	0	0	1	0	0	0	1
NED	0	1	0	1	0	0	1	0	0	1	0	0	1	0
KARL	0	1	0	0	1	0	0	1	0	1	0	0	1	0
KEN	0	1	1	0	0	0	0	1	1	0	0	0	0	1
EARL	0	1	0	0	1	0	0	1	0	1	0	0	0	1
RICK	0	1	0	1	0	0	0	1	0	0	1	0	0	1
OL	0	1	0	1	0	0	1	0	0	1	0	1	0	0
NEAL	0	1	0	1	0	0	0	1	1	0	0	0	1	0
DAVE	0	1	0	1	0	0	0	1	0	0	1	1	0	0

Table 5.9 Elementary statistics of the Gang dataset

Variable	Jet	Shark	Jet %	Shark %
20s	9	1	60.00	8.33
30s	4	9	26.67	75.00
40s	2	2	13.33	16.67
JH	9	1	60.00	8.33
HS	4	7	26.67	58.33
COL	2	4	13.33	33.33
Single	9	4	60.00	33.33
Married	4	6	26.67	50.00
Divorced	2	2	13.33	16.67
Pusher	5	4	33.33	33.33
Bookie	5	4	33.33	33.33
Burglar	5	4	33.33	33.33

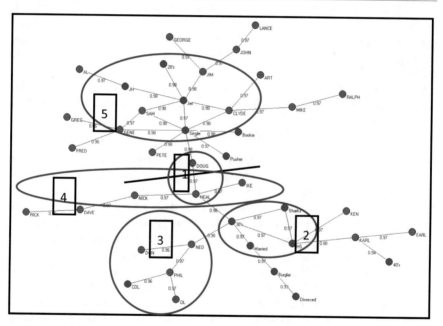

Fig. 5.7 The MRG of the Auto-CM HGC matrix of weights (Red = attributes, Blue = gang's members; Black line = correct separation between Jets and Sharks members)

e. While the borderline Sharks members are grouped in Section 4;

f. The prototypical attributes and members of the Jet gang are clustered together in Section 5, while the Jet borderlines are located outside of the circle of Section 5.

Also in this case the GHC algorithm shows a weighted map of the dataset very useful to understand the multidimensional similarities of attributes and records simultaneously.

Example 4: Attacks against allied forces in Afghanistan

Table 5.10 shows fifty of the more important attacks against the Allied Forces in Afghanistan up to May 2009. Each attack is defined by:

a. The tribe to which the attacker belongs;
b. The ethnic group to which the tribe belongs;
c. And the place where the attack occurred (Latitude and Longitude was not used in this application).

Another dataset defines for each tribe involved in the attacks, an estimation of its military forces (Table 5.11).

We put together the two types of information to build up a data matrix of Boolean values, preprocessed in a suitable way, where each row represents one attack and each column represents different military information about its author, (see Table 5.12 as example).

The key information to extract from this dataset is quite clear; which attacks are connected to each other, which tribes are most often involved in attacks and which tribes are occasionally involved. This task is a particularly suitable for the GHC algorithm since GHC connects in one weighted graph the attacks (rows) and the tribes' military attributes (columns).

Table 5.13 shows the number of attacks that are initiated by some people from the tribes, while Fig. 5.3 shows the MRG executed by GHC after the application of the algorithm to the dataset.

Some observations are necessary about this MRG:

a. Each tribe is positioned in the graph closest to its own typical armaments and closest to the specific attacks committed except for the tribe named "Jadran". We will explain this exception later.
b. The clusters of the attacks are mirrored in their spatial distribution, on the Afghanistan map, even if the information about latitude and longitude were not present in the GHC processing. The algorithm seems to "understand" the "logic" of the attacks from the little information it has about the tribes (see Table 5.14).
c. The GHC and the MRG suggest that 76% of the attacks are committed only by four tribes: Durrani, Jaji, Ghilzai and Safi. This also suggests that the six attacks carried out by the Jadran tribe can be attributed to a strong influence of the Ghilzai tribe on the Jadran tribe. The algorithm also suggests a strong linkage between the Durrani and the Jaji strategy of attacks, while it outlines an independent strategy of the Safi tribe.
d. GHC clusters the other tribes outside a planned strategy of terrorist attacks: the Dari tribe does not seem equipped for systematic attacks while the Panjsheri tribe is the author of only two attacks.
e. The algorithm also suggests an association between the Wardak and Samangani tribes, maybe because their six attacks happened very close to each other in space and in time. However, the algorithm has no information about the location and the sequence of the attacks.

Table 5.10 The 50 most important attacks against the Allied forces in Afghanistan up to May 2009 in time series (confidential source)

Attack	Lat	Lon	Tribe	Ethnic group
1	N34° 22' 26,04"	E62° 10' 26,12"	Dari	Persiani
2	N33° 20' 28,58"	E62° 38' 48,85"	Dari	Persiani
3	N33° 20' 28,55"	E62° 38' 48,60"	Dari	Persiani
4	N33° 20' 29,15"	E62° 38' 47,55"	Dari	Persiani
5	N31° 35' 13,82"	E64° 19' 13,44"	Durrani	Pashtun
6	N32° 21' 16,22"	E63° 23' 58,74"	Durrani	Pashtun
7	N31° 35' 13,40"	E64° 19' 33,84"	Durrani	Pashtun
8	N31° 35' 14,22"	E64° 19' 25,65"	Durrani	Pashtun
9	N31° 35' 14,50"	E64° 19' 36,20"	Durrani	Pashtun
10	N31° 35' 13,58"	E64° 19' 03,31"	Durrani	Pashtun
11	N31° 35' 13,71"	E64° 19' 02,48"	Durrani	Pashtun
12	N31° 35' 13,16"	E64° 19' 12,14"	Durrani	Pashtun
13	N31° 35' 13,33"	E64° 19' 43,27"	Durrani	Pashtun
14	N32° 37' 09,92"	E65° 52' 33,23"	Jaji	Pashtun
15	N32° 37' 09,90"	E65° 52' 34,19"	Jaji	Pashtun
16	N32° 37' 08,85"	E65° 52' 32,66"	Jaji	Pashtun
17	N32° 37' 09,14"	E65° 52' 33,11"	Jaji	Pashtun
18	N32° 37' 09,66"	E65° 52' 33,49"	Jaji	Pashtun
19	N32° 37' 08,79"	E65° 52' 34,20"	Jadran	Pashtun
20	N32° 37' 09,15"	E65° 52' 34,97"	Jaji	Pashtun
21	N32° 37' 09,99"	E65° 52' 33,15"	Jaji	Pashtun
22	N32° 37' 09,27"	E65° 52' 33,55"	Jaji	Pashtun
23	N32° 37' 09,61"	E65° 52' 32,67"	Jadran	Pashtun
24	N32° 37' 09,52"	E65° 52' 35,43"	Jadran	Pashtun
25	N32° 37' 09,41"	E65° 52' 33,76"	Jadran	Pashtun
26	N32° 37' 09,19"	E65° 52' 32,16"	Jaji	Pashtun
27	N32° 37' 10,05"	E65° 52' 32,91"	Jaji	Pashtun
28	N32° 37' 09,10"	E65° 52' 33,48"	Jaji	Pashtun
29	N32° 37' 09,12"	E65° 52' 33,34"	Jaji	Pashtun
30	N33° 27' 03,59"	E69° 44' 26,19"	Safi	Nuristan
31	N33° 27' 37,15"	E69° 44' 13,89"	Jadran	Nuristan
32	N33° 21' 34,55"	E69° 44' 54,71"	Safi	Nuristan
33	N33° 31' 53,21"	E69° 44' 56,26"	Jadran	Nuristan
34	N33° 24' 30,11"	E69° 45' 32,43"	Safi	Nuristan
35	N33° 27' 31,34"	E69° 44' 54,76"	Safi	Nuristan
36	N33° 27' 36,51"	E69° 43' 43,39"	Safi	Nuristan
37	N33° 34' 23,98"	E69° 44' 33,87"	Safi	Nuristan
38	N33° 31' 35,18"	E69° 44' 01,54"	Safi	Nuristan
39	N33° 22465,66"	E69° 44' 26,17"	Safi	Nuristan
40	N34° 15' 46,27"	E70° 49' 05,29"	Ghilzai	Nuristan
41	N35° 51' 46,71"	E71° 38' 16,76"	Panjsheri	Nuristan
42	N34° 22' 46,16"	E71° 55' 43,87"	Samangani	Nuristan
43	N34° 43' 46,31"	E70° 08' 44,23"	Ghilzai	Nuristan
44	N35° 25' 46,77"	E70° 53' 46,59"	Ghilzai	Nuristan
45	N35° 37' 46,65"	E70° 02' 51,32"	Wardak	Nuristan
46	N35° 07' 49,81"	E72° 41' 33,66"	Panjsheri	Nuristan
47	N34° 02' 46,24"	E70° 15' 45,88"	Wardak	Nuristan
48	N34° 28' 47,60"	E70° 37' 36,59"	Wardak	Nuristan
49	N34° 44' 56,27"	E70° 47' 45,53"	Samangani	Nuristan
50	N34° 35' 43,07"	E70° 29' 06,69"	Samangani	Nuristan

Table 5.11 Estimation of the military force of each tribe who to which the attacker belongs (confidential source)

Tribe	Ethnic group	Militants	Light arms			Heavy arms		Vehicles			Vehicles			Mines		C4/TNT (ton)
			Kalashnikov	Stinger	Machine Gun	Tank	Armored vehicles	Jeep	Truck	Motor-bike	Warplane	Helicopter	UAV	Anti Tank Bomb	Anti Men Bomb	
Durrani	Pashtun	1000	1500	50	30	10	5	25	50	100	1	1	3	1000	2500	2.0
Ghilzai	Pashtun	800	1000	50	50	3	10	30	30	60	0	1	0	500	800	0.5
Jadran	Pashtun	650	1000	0	10	5	3	20	25	10	0	0	0	350	200	1.0
Jaji	Pashtun	1200	2000	100	100	20	10	10	60	150	1	2	0	600	2000	3.5
Khugian	Pashtun	800	900	30	10	2	0	15	15	0	0	0	0	300	450	0.5
Mangal	Pashtun	900	1100	80	30	5	1	22	28	13	0	0	0	150	600	1.0
Mohammed	Pashtun	250	300	0	30	0	0	3	5	10	0	0	0	50	100	0.0
Safi	Pashtun	1300	1500	40	100	6	2	25	35	95	2	2	1	500	750	4.0
Shinwari	Pashtun	300	400	10	5	0	0	2	8	13	0	0	0	20	100	0.5
Tani	Pashtun	450	500	20	5	0	0	3	11	45	0	0	0	40	150	0.5
Wardak	Pashtun	350	400	10	10	1	2	4	7	15	0	0	0	100	150	0.0
Panjsheri	Tajiks	400	500	50	10	0	1	2	20	35	0	0	0	150	200	0.0
Andarabi	Tajiks	500	550	0	20	0	0	5	30	28	0	0	0	170	200	1.0
Samangani	Tajiks	350	400	10	20	0	0	1	14	12	0	0	0	30	60	0.0
Badakhshi	Tajiks	450	600	30	40	0	1	5	26	30	0	0	0	100	100	0.0

Table 5.12 An example of the data matrix we intend to process

	Tribe Dari	Tribe Durrani	Tribe Ghilzai	Tribe Jadran	Tribe Jaji	...	Tribe Wardak	Nuristan	Pashtun	Persian	Militants(350)	Militants(400)	...	C4/ton
A1	1	0	0	0	0	0	0	1	0	0	0	0	0	0
A2	1	0	0	0	0	0	0	1	0	0	0	0	0	0
A3	1	0	0	0	0	0	0	1	0	0	0	0	0	0
A4	1	0	0	0	0	0	0	1	0	0	0	0	0	0
A5	0	1	0	0	0	0	0	0	1	0	0	0	0	0
A6	0	1	0	0	0	0	0	0	1	0	0	0	0	0
A7	0	1	0	0	0	0	0	0	1	0	0	0	0	0
A8	0	1	0	0	0	0	0	0	1	0	0	0	0	0
A9	0	1	0	0	0	0	0	0	1	0	0	0	0	0
A10	0	1	0	0	0	0	0	0	1	0	0	0	0	0
A11	0	1	0	0	0	0	0	0	1	0	0	0	0	0
A12	0	1	0	0	0	0	0	0	1	0	0	0	0	0
A13	0	1	0	0	0	0	0	0	1	0	0	0	0	0
A14	0	0	0	0	1	0	0	0	0	0	0	0	0	0
A15	0	0	0	0	1	0	0	0	0	0	0	0	0	0
A16	0	0	0	0	1	0	0	0	0	0	0	0	0	0
A17	0	0	0	0	1	0	0	0	0	0	0	0	0	0
A18	0	0	0	1	0	0	0	0	0	0	0	0	0	0
A19	0	0	0	0	0	0	0	0	0	0	0	0	1	0
A20	0	0	0	0	1	0	0	0	0	0	0	0	0	0
A21	0	0	0	0	1	0	0	0	0	0	0	0	0	0
A22	0	0	0	1	0	0	0	0	0	0	0	0	0	0
...	0	0	0	1	0	0	0	0	0	0	0	0	1	0
A50	0	0	0	1	0	0	0	0	0	0	0	0	1	0

Table 5.13 The number of attacks executed by the tribes

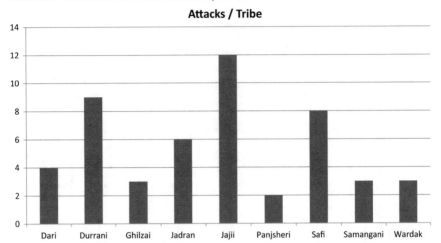

f. It is interesting that the first attack committed by the Durrani tribe (A5) is indicated by GHC as the hub of all attacks of this tribe. In the same way, the first attack committed by Jaji tribe (A14) is again the hub of all attacks of Jaji tribe. But remember that the dataset processed did not include temporal information.

g. The tribes in the south of the graph (Dari, Panjsheri, Jadran, Samangani and Wadrak) have fewer armaments than the other four tribes (Durrani, Jaji, Safi and Ghilzai). Therefore, the algorithm split the graph into two fuzzy regions: the area of the most dangerous tribes, and the area of the tribes whose attacks are less organized (Fig. 5.8).

5.3.4   Discussion of GHC

We have shown that the GHC algorithm is able to detect the complex and hidden relations between records and variables in different data sets. The complexity of the Rooms data set arises from the high number of the variables, fifty, compared to the small number of the records, only six. All the same, the GHC algorithm depicts a framework where each room is defined by its correct attribute. The global topology of the graph is a mirror of the similarity of the different rooms. The same task would also be hard to perform by a human using his/her contextual knowledge about the data set.

The Gang data set is an intriguing artificial sample of a situation where attributes are spread out into the records in a tricky way. For this reason this data set was used in other scientific efforts about content addressable memory [2, 12]. The last data set is a sample of real 50 terrorist attacks that happened in Afghanistan up to May

Table 5.14 Comparison between the real authors of the attacks and the attacks responsibilities according to the GHC graph (in bold the discordances)

Attacks #	Long	Lat	Graph tribe	Real Tribe	Attacks #	Long	Lat	Graph tribe	Real Tribe
A1	62.17392222	34.3739	Dari	Dari	A26	65.8756	32.61921944	Jaji	Jaji
A2	62.64690278	33.34127222	Dari	Dari	A27	65.87580833	32.61945833	Jaji	Jaji
A3	62.64683333	33.34126389	Dari	Dari	A28	65.87596667	32.6191944	Jaji	Jaji
A4	62.64654167	33.34143056	Dari	Dari	A29	65.87592778	32.6192	Jaji	Jaji
A5	64.3204	31.58717222	Durrani	Durrani	A30	69.74060833	33.45099722	Safi	Safi
A6	63.39965	32.35450556	Durrani	Durrani	A31	69.73719167	33.46031944	Ghilzai	Jadran
A7	64.32606667	31.58705556	Durrani	Durrani	A32	69.74853056	33.35959722	Safi	Safi
A8	64.32379167	31.58728333	Durrani	Durrani	A33	69.74896111	33.53144722	Ghilzai	Jadran
A9	64.32672222	31.58736111	Durrani	Durrani	A34	69.75900833	33.40836389	Safi	Safi
A10	64.31758611	31.58710556	Durrani	Durrani	A35	69.74854444	33.45870556	Safi	Safi
A11	64.31735556	31.58714167	Durrani	Durrani	A36	69.72871944	33.46014167	Safi	Safi
A12	64.32003889	31.58698889	Durrani	Durrani	A37	69.74274167	33.57332778	Safi	Safi
A13	64.32868611	31.58703611	Durrani	Durrani	A38	69.73376111	33.52643889	Safi	Safi

(continued)

Table 5.14 (continued)

Attacks #	Long	Lat	Graph tribe	Real Tribe	Attacks #	Long	Lat	Graph tribe	Real Tribe
A14	65.87589722	32.61942222	Jaji	Jaji	A39	69.74060278	33.38490556	Safi	Safi
A15	65.87616389	32.61941667	Jaji	Jaji	A40	70.81813611	34.26285278	Ghilzai	Ghilzai
A16	65.87573889	32.619125	Jaji	Jaji	A41	71.63798889	35.862975	Panjsheri	Panjsheri
A17	65.87586389	32.61920556	Jaji	Jaji	A42	71.92885278	34.37948889	Samangani-Wardak	Samangani
A18	65.87596944	32.61935	Jaji	Jaji	A43	70.14561944	34.72953056	Ghilzai	Ghilzai
A19	65.87616667	32.61910833	Ghilzai	Jadran	A44	70.896275	35.42965833	Ghilzai	Ghilzai
A20	65.87638056	32.61920833	Jaji	Jaji	A45	70.04758889	35.629625	Samangani-Wardak	Wardak
A21	65.875875	32.61944167	Jaji	Jaji	A46	72.69268333	35.13050278	Panjsheri	Panjsheri
A22	65.87598611	32.61924167	Jaji	Jaji	A47	70.26274444	34.04617778	Samangani-Wardak	Wardak
A23	65.87574167	32.61933611	Ghilzai	Jadran	A48	70.62683056	34.47988889	Samangani-Wardak	Wardak
A24	65.87650833	32.61931111	Ghilzai	Jadran	A49	70.79598056	34.74896389	Samangani-Wardak	Samangani
A25	65.87604444	32.61928056	Ghilzai	Jadran	A50	70.48519167	34.59529722	Samangani-Wardak	Samangani

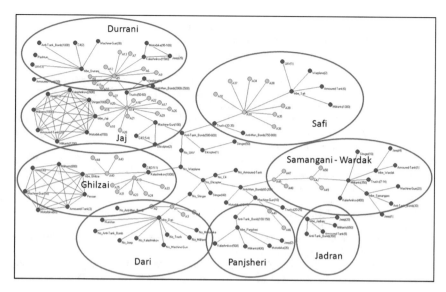

Fig. 5.8 MRG of the GHC algorithm results

2009. The graph detected by the GHC algorithm reflects many of the analyses of the experts of this field [13–15].

References

1. Buscema, M., V. Consonni, D. Ballabio, A. Mauri, G. Massini, M. Breda, and R. Todeschini. 2014. K-CM: A New Artificial Neural Network. Application to Supervised Pattern Recognition. *Chemometrics and Intelligent Laboratory Systems* 138: 110–119.
2. McCLelland, J.L., D.E. Rumelhart, and G.E. Hinton. 1986. The Appeal of Parallel Distributed Processing. In *Parallel Distributed Processing*, Chap. 1 vol. 1, ed. J.L. McClelland and D.E. Rumelhart, 3–44. Exploration in Microstructure of Cognition, Foundations, MIT Press.
3. Buscema, M. 2007. A Novel Adapting Mapping Method for Emergent Properties Discovery in Data Bases: Experience in Medical Field. In *2007 IEEE International Conference on Systems, Man and Cybernetics (SMC 2007)*. Montreal, Canada, October 7–10, 2007.
4. Buscema, M., and E. Grossi. 2008. *The Semantic Connectivity Map: An Adapting Self-Organizing Knowledge Discovery Method in Data Bases. Experience in Gastro-oesophageal reflux disease.* International Journal Data Mining and Bioinformatics 2 (4) 2008.
5. Buscema, M., F. Newman, and E. Grossi, W. Tastle. 2010. Application of Adaptive Systems Methodology to Radiotherapy. In *NAFIPS*, July 12–14, 2010 Toronto, Canada.
6. Di Ludovico, A., and G. Pieri. 2011. Artificial Neural Networks and Ancient Artefacts: Justifications for a Multiform Integrated Approach Using PST and Auto-CM Models. *Archeologia e Calcolatori* 22: 99–128.
7. Coppedè, F., E. Grossi, M. Buscema, and L. Migliore. 2013. Application of Artificial Neural Networks to Investigate One-Carbon Metabolism in Alzheimer's Disease and Healthy Matched Individuals. PLOS ONE, 8 (8): e74012, 1–11.

8. Gironi, M., B. Borgiani, E. Farina, E. Mariani, C. Cursano, M. Alberoni, R. Nemni, G. Comi, M. Buscema, R. Furlan, and Enzo Grossi. 2015. A Global Immune Deficit in Alzheimer's Disease and Mild Cognitive Impairment Disclosed by a Novel Data Mining Process. *Journal of Alzheimer's Disease* 43: 1199–1213.

9. F. Drenos, E. Grossi, M. Buscema, and S. E. Humphries, *Networks in Coronary Heart Disease Genetics As a Step towards Systems Epidemiology*, PLoS ONE 10(5): May 7 (2015). e0125876. https://doi.org/10.1371/journal.pone.0125876.

10. Coppedè, F., E. Grossi, A. Lopomo, and R. Spisni. 2015. M Buscema & Lucia Migliore. *Application of Artificial Neural Networks to Link Genetic and Environmental Factors to DNA Methylation in Colorectal Cancer, Epigenomics* 7 (2): 175–186.

11. Buscema, M., E. Grossi, L. Montanini, and M. E. Street. 2015. *Data Mining of Determinants of Intrauterine Growth Retardation Revisited Using Novel Algorithms Generating Semantic Maps and Prototypical Discriminating Variable Profiles*, PLoS ONE 10(7): June 9, 2015 e0126020. https://doi.org/10.1371/journal.

12. McClelland, J.L., and D.E. Rumelhart. 1988. *Explorations in Parallel Distributed Processing*, Chap. 2, 11–47. MIT Press.

13. Tribal Analysis Center. 2009. Ishaqzai Tribe, Tribal Analysis Center, 6610-M Mooretown Road, Box 159. Williamsburg, VA, 23188

14. Thomas, Ruttig. 2009. The Other Side: Dimensions of the Afghan Insurgency: Causes, Actors and Approaches to Talks, Kabul/Berlin: Afghanistan Analysts Network, Thematic Report 01/2009, July 2009.

15. Schubert, Janis L., W. Thomas Moore, Patrick D. Finley, Ryan Hammer, and Robert J. Glass. 2010. *Modeling Tribal Leadership Dynamics: An Opinion Dynamics Model of Pashtun Leadership Selection*. Sandia National Laboratories.

16. Buscema, M., and P. L. Sacco. 2010. Auto-contractive Maps, The H Function, and The Maximally Regular Graph (MRG): A New Methodology for Data Mining. In *Applications of Mathematics in Models, Artificial Neural Networks and Arts*, Chap. 11, eds.V. Capecchi et al. https://doi.org/10.1007/978-90-481-8581-8_11. Springer.

17. Buscema, M., M. Asadi-Zeydabadi, W. Lodwick, and M. Breda. 2016. The H_0 Function, A New Index for Detecting Structural/Topological Complexity Information in Undirected Graphs. *Physica A* 447: 355–378.

Chapter 6
Advances, the K-Contractive Map: A Supervised Version of Auto-CM

Abstract This section is devoted to a more advanced type of Auto-CM that is supervised.

6.1 Introduction

There are over 100 different types of ANNs in use today which indicates that there are a variety of ways to approach ANNs and their efficacy.

A common problem with ANNs is that many researchers consider ANNs as "black boxes," [1]. In particular, determining why an ANN makes a particular decision is a difficult task. This is a significant weakness, for without the ability to produce comprehensible decisions it is hard to trust the reliability of networks addressing actual problems. In other words many people refuse to use ANNs because there is no satisfactory interpretation of their behavior. ANNs in fact capture "hidden" relations between inputs and outputs with a high degree of accuracy, but do not indicate why the particular resulting structure works. The usual mathematical modeling approach that "imposes" a mathematical structure to the underlying problem does not suffer from "hidden" relations since its structure is determined a priori and explicitly chosen for proscribed and known reasons.

The information stored in a neural network is a set of numerical weights and connections that provides no direct clues as to how the task is performed or how the relationships between inputs and outputs were derived. This limits the usage and acceptance of ANNs since, in many applications in science and engineering, techniques must be based on analytical functions that can be understood. Further complicating the use of ANNs is the tedious process of parameter selection. Even when performing very similar tasks, the proper choice of network parameters can vary widely. These parameters, which include the neural network structure (architecture, number of layers and number of nodes per layer), error bounds, learning rates, training algorithms, hidden layer sizes and the data vector used, are often chosen in a trial-and-error process.

Another problem lies in the initialization process. If the start is from random weights on the connections which is often the case, every resulting ANN will

© Springer International Publishing AG, part of Springer Nature 2018

P. M. Buscema et al., *Artificial Adaptive Systems Using Auto Contractive Maps*, Studies in Systems, Decision and Control 131, https://doi.org/10.1007/978-3-319-75049-1_6

represent a unique conceptualization of the problem under study. That is, the neural network will potentially end up with a different set of weights that represent different models derived from the different random starts. On the other hand, obtaining different models (sets of weights) from different initializations has some advantages in certain situations. When different resulting models occur from the various random starts, it means that the underlying problem is more complex than the structure of the ANN is able to capture. It also implies that there is a need to assemble many runs of independent trials to understand the complexity of the underlying structure. The variation of the results will indicate the risk of implementing final results. That is, varying the final neural network weights that produce the varying results indicate instability of the neural network model.

All these problems have prompted some researchers such as ourselves to develop a new conceptual type of supervised ANNs [2] with the purpose of overcoming some of the limitations of current ANNs. In particular, a new type of ANN must address at least the following:

- The need for random initialization for the given dataset to determine how the ANN structure performs and in this way to determine the stability of the design;
- The need to detect over-fitting and stop the process when this occurs;
- The need to provide to the user a conceptual map (an acyclic directed weighted graph) that shows how its variables are organized after the learning phase;
- The need to provide the user with a conceptual map (undirected weighted graph) that shows the similarities among all the testing patterns.

K-Contractive Map (K-CM) is a supervised artificial neural network that is capable of supporting its decisions by using:

- A weighted semantic graph of records that are evaluated in blind testing;
- A semantic graph that shows how the network organized the relations between all the attributes.

6.2   K-CM Algorithm

K-CM uses the following tools:

1. *Auto-CM* that was presented in Chap. 3;
2. The FS-transform algorithm presented in Chap. 5, that will equip us with specific equations through which each variable of each record in the training data set and each variable of each record from the testing data set are redefined in terms of a fuzzy membership grade. This fuzzy membership is determined on the basis of the connections matrix necessary to complete the grid generated by the Auto-CM system at the end of the training phase;
3. An algorithm to determine the nearest-neighbor for the assignment of a class to each record;

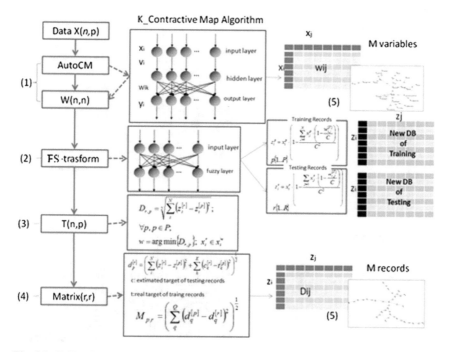

Fig. 6.1 K-CM algorithm processing steps

4. An algorithm to organize specific distance matrices derived from the training and testing records and redefined by the FS-transform algorithm;
5. An algorithm to find the distance matrices in an un-weighted directed graph, the MRG, presented in Chap. 4.

Figure 6.1 illustrates these five steps of the K-CM processing algorithm.

The K-CM approach provides a "bottom-up" symbolic explanation of its learning and behavior. At the symbolic level, K-CM is not, however, a set of naïve "If … Then" rules traditionally applied to all training records. This symbolic level of understanding is not what we seek. That is, we do not seek a nonlinear function that interpolates the training set results defined by a set of explicit rules suitable to model the underlying process. What we seek is a more interesting symbolic level approach, one that allows the explanation of the "fuzzy mental map" through which the learning algorithm is represented on the basis of training data and, simultaneously, the subjective similarities on which the same algorithm can operate on new cases. In other words, we seek a bottom-up algorithm that can work in a symbolic way and is able to self-generate a mental representation (weighted graph) of what is learned (training set) and what dynamically took place represented in a chart of new experiences (test set). K-CM is a hybrid system, processing all the source information (patterns) by means of different steps, each one of which provides an

Fig. 6.2 Architecture of the
Auto-CM

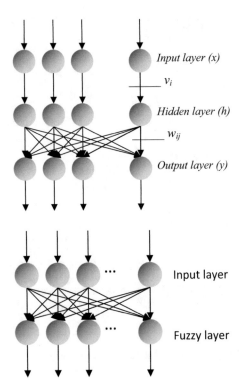

Fig. 6.3 Architecture of the
K-CM

interpretation of the previous one. This methodology is known in the literature as
the Restricted Boltzmann Machine (RBM) and is considered deep learning [3–5].

Step 1: The Application of Auto-CM
Auto-CM was presented in Chap. 3. The reader is directed to that chapter for assistance in understanding of this section if there is a need. Figure 6.1 can serve as a reminder of Auto-CM. This is the first step of the K-CM algorithm. That is, the first step in the K-CM algorithm applies Auto-CM to the given dataset (Fig. 6.2).

Step 2: Creation of The Fuzzy Dataset
The architecture associated with Step 2 of the K-CM algorithm employs a two layer network as illustrated by Fig. 6.3, according to the FS-transform equation restated below. This involves the input values in the training phase, of the fuzzified weights of the full grid connections matrix generated by the Auto-CM system at the end of the training phase.

$$z_i^{[p]} = x_i^{[p]} \left(1 - \frac{\sum_{j=1}^{N} x_j^{[p]} \cdot \left(1 - \frac{w_{i,j}^{[T]}}{C} \right)}{C} \right) \tag{6.1}$$

This transformation generates a new dataset where the original values of the training dataset are reformulated in a fuzzy way (see Chap. 5).

Step 3: Testing Phase

A new record from the FS-transform equation is typed with the patterns of a known class membership (the initial dataset redefined in a fuzzy way) and used together with others to determine its own class. At this stage, the given new non-classified record, x_i^{R+1}, has a fuzzy (gradient) value

$$z_i^{R+1} = x_i^{R+1} \left(1 - \frac{\sum_{j=1}^{N} x_j^{R+1} \cdot \left(1 - \frac{w_{i,j}^{[T]}}{C} \right)}{C} \right) \tag{6.2}$$

where C comes from Auto-CM (see Chap. 3 and FS-transform of Chap. 5). The new record contributes to its own classification by using the distance of the new record from those classified in the learning phase. The output strength of each trained weight.

$$F_i = \sum_{j}^{N} w_{i,j}^{[T]} \tag{6.3}$$

weighs the distance between one record of testing and all the training records,

$$D_{R+1,p} = \sqrt{\sum_{i=1}^{N} F_i \cdot \left(z_i^{[p]} - z_i^{R+1} \right)^2} \quad \forall p \in R \tag{6.4}$$

and computing

$$w = \mathrm{argminin}_p \left\{ D_{R+1,p} \right\} \tag{6.5}$$

the result of the classification will be,

$$x_i^{R+1} \in x_i^{w} \tag{6.6}$$

Step 4: The Construction of the Graph

The K-CM algorithm, once trained on the training sample, is in a position to represent (utilizing a semantic map) the values of the input vector (typing) and suggests a classification (recognition) scheme. The specific algorithm by which K-CM carries out this projection is called the *Meta-Distance Algorithm*. In formal terms, we calculate the Euclidean distance of each test record (input + assumed target) from all records used for training, input + real targets using the following equation

$$d_q^{[p]} = \left(\sum_{i}^{N} \left(x_i^{[p]} - y_i^{[q]} \right)^2 - \sum_{k}^{K} \left(x_i^{[p]} - y_i^{[q]} \right)^2 \right)^{\frac{1}{2}} \tag{6.7}$$

Table 6.1 Notation for the K-CM algorithm

Symbol	Meaning
$i \in N$	where N is the number of input attributes
$k \in K$	where K is the number of output classes
$p, r \in P$	where P is the number of testing records
$q \in Q$	where Q is the number of training records
$x_i^{[p]}$	ith input attribute of the pth record in testing, after application of the K-CM FS-transform equations
$y_i^{[q]}$	ith input attribute of the qth record in training, after application of the K-CM FS-transform equations
$c_k^{[p]}$	kth class to which belongs the estimated pth record in testing (real targets, 1 of N)
$t_k^{[q]}$	kth class which really belongs to the qth record in testing (real targets, 1 of N)
$d_q^{[p]}$	Euclidean distance of the pth record from the qth record in training

$$M_{p,r} = \left(\sum_q^Q \left(d_q^{[p]} - d_q^{[r]} \right)^2 \right)^{\frac{1}{2}} \tag{6.8}$$

The notation for the K-CM algorithm that are associated with Eq. (6.7) are given in Table 6.1.

The result of this operation generates a new P by Q matrix where the rows represent the test records and the columns represent the training records. This matrix, which for convenience we denote R, is a new dataset which measures how distant, from the point of view of K-CM, all test records are from each training record. It then computes the Euclidean distance between each test record with each other starting from this new matrix R of relative distances, according to Eq. (6.7) above. We define a *Meta Distance (M)* matrix as this new distance matrix between the test records based on the new matrix R as follows,

Step 5: The Matrix M is Displayed Using the MST

Once the matrix M is computed, the MST is used to display the results. We present the five examples where we explain how we prepared each example according to steps 1–4 outlined above. Other examples can be found in [2] and in [6]. Moreover, we show how we split the dataset into the training and testing sub-datasets. That is, for each of the five tests we performed below, we show how we processed each of the five datasets and how the K-CM algorithm produced a complete array of weights. This array represents the knowledge learned from the K-CM algorithm on each of the specific subsets of training. The five weight matrices generated for each dataset below in our examples had a correlation of over 98%. This indicates that the five training subsets created from the K-CM algorithm contains roughly the same knowledge. The sum of five weight matrices, then, for each of the five datasets, should represent the global knowledge of K-CM algorithm with respect to the dataset. The MRG of this global matrix is the "weighted mind map" with which the K-CM system represents each of the five datasets (all these datasets are present in UCI Machine Learning Repository at https://archive.ics.uci.edu/ml/datasets.html).

6.3 Examples

6.3.1 Iris Dataset

This dataset consists of 150 records, classifying the Iris varieties into three different classes (Silky, Virginica and Versicolor) starting with four numeric attributes that represent the size of the petals and sepals. The distribution of grades is divided into 50 records for each class. The dataset is structured to separate linear records that identify the Silky from the other two types of irises, while the Virginica and Versicolor overlap. Figure 6.4 shows the correlation between the independent variables, as identified by K-CM, with respect to the Iris dataset. The graph shows a strong correlation between the values of the size of the petals as demonstrated by how the dataset is strongly dependent on the size of the sepals (length, width) and less dependent on one of the petals. This is probably due to the properties of the dataset itself which has a pronounced separation between the Silky Variety and the other two varieties of irises, while between the Virginica and Versicolor there is a partial overlap. Therefore, the size of the sepals is critical to differentiating the Silky from the Virginica and Versicolor, while the size of the petals provides us with the information necessary to classify the varieties of iris that overlap. To confirm this we conducted an experiment through the Multilayer Perceptron of Weka (https://weka.waikato.ac.nz) with only two variables describing the sepals. The network has achieved an 82% overall accuracy, committing only one error on Silky class and 19 errors on the remaining two classes. By adding one of the two variables describing the petal (width or length) global accuracy rises to 95% by committing errors (5) only on the recognition of the Virginica class.

6.3.2 Digits Dataset

This dataset is composed 5620 records and 64 independent variables which represent the handwritten digits from 43 different people on a pre-printed form. The pre-printed forms were sampled in a 32×32 pixel bitmap. Each bitmap was divided into blocks of 4×4 pixels forming an 8×8 matrix which contains the number of pixels in each block by 4×4. This procedure generates 64 dataset attributes. There are 10 dependent variables and correspond to numeric digits 0–9. The graph of digital dataset represents the dataset of the optical recognition of handwritten digits, showing the

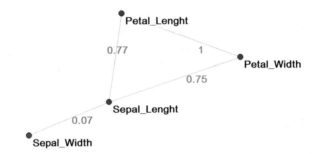

Fig. 6.4 Iris-MRG of the independent variables

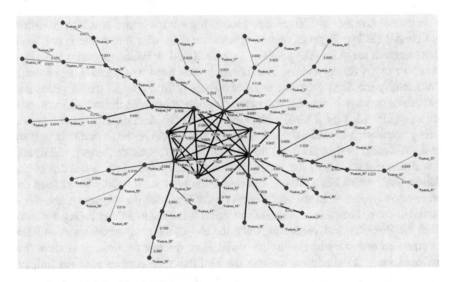

Fig. 6.5 Digits-MRG of independent variables

presence of a significant cluster at the center of the map. This mini-graph illustrates how K-CM groups variables (features) that have a frequency of pixels constant for all patterns (Fig. 6.5).

6.3.3 Sonar Dataset

This dataset represents the classification of sonar signals that are reflected on a cylindrical metal surface or on a rock approximately cylindrical in shape. Each record is composed of 60 numerical attributes between 0.0 and 1.0, representing the intensity of sonar signal reflection. The target is specified if the obstacle encountered is a mine or a rock. The dataset consists of a total of 208 records, 111 mine records and 97 rock records. The dataset Sonar (Fig. 6.6) shows, as in the case of the digits

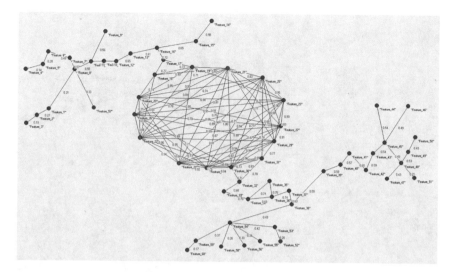

Fig. 6.6 Sonar-MRG of independent variables

example, a significant central cluster in which certain features have a significantly higher correlation than all other variables. In this case, interestingly, K-CM arranges variables that describe the sonar signals to uniform frequencies from which sonar signals were generated.

6.3.4 Steel Plates Defects Dataset

This dataset classifies the seven different types of defects found in the production of steel plates (Pastry, Z_Scratch, K_Scatch, Stains, Dirtiness, Bumps, and Other_Faults). Each record in the dataset is described by 27 attributes that describe the shape and appearance of defects for a total of 1941. This case is one which K-CM reorganizes the variables into two distinct clusters (Fig. 6.7). At the bottom are the poorly correlated variables while the other cluster shows how variables are organized around the "Edge_Y_Index" attribute forming a small graph. The MRG generated by K-CM shows how this particular dataset is very complex and difficult to classify.

6.3.5 Vowels Dataset

This dataset consists of 990 records representing the pronunciation of 11 English vowels. Each vowel has been associated with a word spoken six times by 15 persons, eight men and seven women. Every word spoken has been sampled and processed

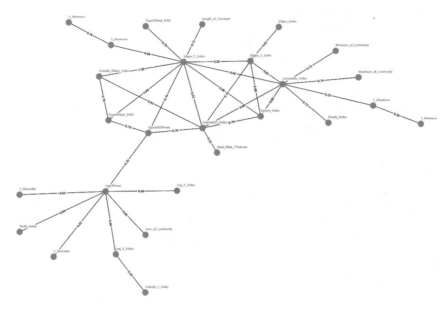

Fig. 6.7 Steel Plates-MRG of independent variables

by 10 independent variables that represent input attributes. There are 11 classes and refer to the vowels pronounced. Originally, the experiment was divided into two distinct datasets, the training set (528 records), produced by the pronunciation of four men and three women. The dataset used here is the composition of these two datasets. K-CM groups the variables into two triangles (Fig. 6.8) joined together by the attribute Feature_9 that is also the center of the graph. Particularly interesting is the upper triangle showing a very strong correlation between these three variables (Features 9-2, 9-3).

6.4 An Example of K-CM Graph of Records: Analysis of the Optical Dataset

We next focus our attention on the Digital Optical dataset that we previously have analyzed from the point of view of the correlation among the variables (see 6.2) in order to show how K-CM is also able to project in a weighted graph the records of a dataset. Recall that the Digital Optical testing dataset, in fact, is composed of 1401 records representing 8×8 pixel handwritten digits in Fig 6.9.

The graph resulting from the K-CM maximal regular graph processing offers a joint analog of the various classes of digits below. For example, it is known that the class of the digit "1" is connected to that of the digit "7" and that of "6" to that of

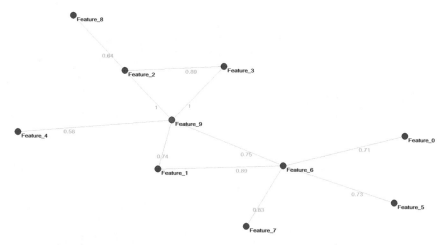

Fig. 6.8 Vowels-MRG of independent variables

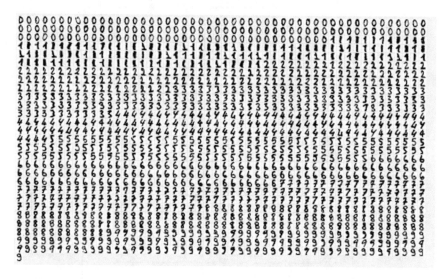

Fig. 6.9 The testing set of 1401 handwritten images of digits each in an 8 × 8 bit size

"5". Furthermore, Table 6.2 is an analysis of the 19 digits to which the K-CM has attributed a wrong target is shown in Fig. 6.10.

One may notice that some of these errors seem to be motivated by the noise of the record, and in particular, record 161 is a "1" but that the network has given it the target "7", the record 543 is a "3" but recognized to be a "1", record 544 is a "3" but is recognized as an "8", record 1159 is an "8" but has been given the target of "9", and finally record 1062 is an "8" but is mistaken for a "1" (Figs. 6.11, 6.12 and 6.13).

Table 6.2 The digits confusion matrix

Confusion matrix	Fig1	Fig2	Fig3	Fig4	Fig5	Fig6	Fig7	Fig8	Fig9	Fig0	Row	Errors	Accuracy %
Figure1	138	0	0	0	0	0	0	0	0	0	138	0	100.00
Figure2	0	141	0	0	0	0	0	1	0	0	142	1	99.30
Figure3	0	0	138	0	0	0	0	0	1	0	139	1	99.28
Figure4	0	1	0	139	0	0	0	0	1	2	143	4	97.20
Figure5	0	0	0	0	142	0	0	0	0	0	142	0	100.00
Figure6	0	0	0	1	0	136	0	0	0	2	139	3	97.84
Figure7	0	0	0	0	0	0	139	0	0	0	139	0	100.00
Figure8	0	0	0	0	0	0	0	140	0	1	141	1	99.29
Figure9	0	1	0	0	0	0	0	0	136	1	138	2	98.55
Figure0	0	1	0	1	2	0	0	2	1	133	140	7	95.00
Col total	138	144	138	141	144	136	139	143	139	139	1401	19	
Arithmetic mean accuracy													98.65
Weighted mean accuracy													98.64

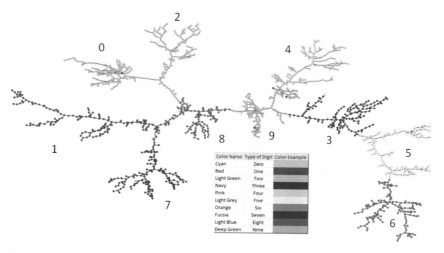

Fig. 6.10 Graph produced by K-CM MRG on the testing set of the dataset digits consisting of 1401 records

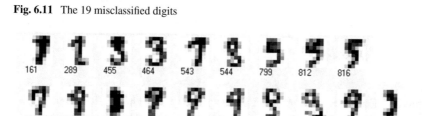

Fig. 6.11 The 19 misclassified digits

Fig. 6.12 The 19 misclassified digits zoomed to 1 × 5 pixels

6.5 Final Considerations

6.5.1 Over Fitting

It is important to understand the concept of over fitting in machine learning. Over fitting occurs when a model learns past the point of generalizing from a learned trend. That is, the model takes the training data, sufficient to identify the trends, and applies that knowledge to the remainder of the data, the test data. As an extreme example we take a data system in which the number of variables is the same as or greater than the number of observations. In this case a simple model or learning process can perfectly predict the training data simply by memorizing the training data in

record	Real digit	K-CM extimation	8 x8 box	Box Zoom
161	1	7	,	
289	2	8	t	
455	3	9	s	
464	3	9	3	
543	3	1	1	
544	3	8	s	
799	5	9	,	
812	5	9	,	
816	5	3	,	

/ ⅃ ⇒

record	Real digit	K-CM extimation	8 x8 box	Box Zoom
1014	7	9	η	
1159	8	9	,	
1062	8	1	,	
1310	9	7	,	
1318	9	7	,	
1333	9	4	1	
1353	9	8	s	
1359	9	3	٦	
1365	9	4	η	
1369	9	1	,	

Fig. 6.13 Table of the errors

its entirety. However, such a model will typically fail dramatically when making predictions about new or unseen data, since the simple model has not learned to generalize. Therefore, it is useful to understand how to make machine learning models in a more robust way. K-CM is not prone to over fitting. This problem is addressed at the beginning of the K-CM algorithm. The K-CM algorithm does not perform a direct minimization of an error function between targets and the ANN output. K-CM works with a complex minimization of the energy function involving all the input variables as a set of parallel constraints. We call this "complex minimization" because K-CM is a supervised ANN, and it first defines the continuous hyper-surface for all the input variables. Then it associates this hyper-surface to the dependent variables (output). The complex minimization of K-CM is a multi-objective minimization. In fact, if we look at the equations of Auto-CM first and then K-CM, we can see that two minimization criteria that work together, one after another: step 1 (Fig. 6.1) where the weights matrix is defined, and step 2 and 3 together (see Fig. 6.1), where the FS-transform and the weighted kNN test are performed. In both cases we can see two energy (error) minimization criteria applied in sequence.

6.5.2 White Box

K-CM output provides also a weighted graph of the test records where the similarity (and so the correlation) among records is shown. The number of subjects and variables can be very high, but despite this complexity, the K-CM algorithm is able to cope with an explosive dimensionality in the observation vector [2]. The MST "filter" where relations among variables with small weights have no edge results in a graphical representation of the inherent structure of the similarity matrix explicitly indicating how closely variables resemble one another. The MRG filter depicts loops of additional complexity in cases where some records have major degrees of similarity. Furthermore, in comparative studies K-CM has shown to have the best classification performance in validation for most of the datasets on which it was tested [2]. On average, K-CM out-performed the other classification methods. Among all the tested methods, only k-NN (learning machine [2]) demonstrated that it has a similar performance as that K-CM, they being based on the same modelling principles. Therefore, it can be concluded that all the methods that exploit the k-NN strategy are reliable methods for classification and the novel K-CM can improve the classification results especially in those cases where nonlinear relationships among variables are relevant.

References

1. Chauvin, Y., and D.E. Rumelhart (eds.). 1995. *Backpropagation: Theory, Architectures, and Applications, Lawrence Erlbaum Associates*. New Jersey: Inc. Publishers.
2. Buscema, M., V. Consonni, D. Ballabio, A. Mauri, G. Massini, M. Breda, and R. Todeschini. 2014. K-CM: A New Artificial Neural Network. *Application to Supervised Pattern Recognition, Chemometrics and Intelligent Laboratory Systems* 138: 110–119.
3. Hinton, G.E., S. Osindero, and Y.-W. Teh. 2006. A Fast Learning Algorithm for Deep Belief Nets. *Neural Computation* 18 (7): 1527–1554.
4. Bengio, Y. 2009. Learning Deep Architectures for AI. *Machine Learning* 2 (1): 1–127.
5. Larochelle, H., and Y. Bengio. 2008. Classification Using Discriminative Restricted Boltzmann Machines. In *Proceedings of the 25-th International Conference on Machine Learning*, Helsinki, Finland.
6. Gironi, M., B. Borgiani, E. Farina, E. Mariani, C. Cursano, M. Alberoni, R. Nemni, G. Comi, M. Buscema, R. Furlan, and Enzo Grossi. 2015. A Global Immune Deficit in Alzheimer's Disease and Mild Cognitive Impairment Disclosed by a Novel Data Mining Process. *Journal of Alzheimer's Disease* 43 (2015): 1199–1213.

Chapter 7
Comparison of Auto-CM to Various Other Data Understanding Approaches

Abstract We compare Auto-CM with various other methods that extract patterns from data. The way that we measure the results of comparisons uses MST.

7.1 Introduction

The availability of data is constantly expanding [1]. This means that the need to sort, order, and understand their basic structure must also expand [2]. Multidimensional scaling techniques are typically called on to take on this role [3]. Among the most widely adopted multi-dimensional scaling techniques is Principal Component Analysis (PCA) [4, 5], thanks to its conceptual simplicity and clarity and ease of interpretation of results. On the other hand, PCA has well known intrinsic limitations in terms of orthogonality, mean-variance representativeness, and scale, and therefore its use is not always advisable, for instance when the inherent statistical model is not known [6].

A clear conceptual roadmap for choosing among different techniques depending on the dataset and statistical process features is currently not available. Therefore what we are left with is to perform comparative tests among different techniques on well-known classes of problems to try to understand which techniques work on which problem/dataset types [7, 8]. We propose a new approach to this problem, by introducing a fitness criterion for technique assessment that is shown to function effectively for a variety of problems. We compare well known techniques such as the aforementioned PCA, and other popular ones like linear correlation and Manhattan similarity, with ANN-based techniques such as the Self-Organizing Map (SOM) [9], and a relatively new one, the Auto-Contractive Map (Auto-CM) [10, 11]. We show that Auto-CM qualifies as the best performer for the whole battery of datasets, and therefore may be of special interest in cases where more conventional techniques fare poorly.

Our comparison measure criterion is built upon the Minimal Spanning Tree (MST) (see Chap. 4 and [12]). The literature provides many examples of successful deployment of MST-based approaches to capture dataset complexity and the ability to extract useful information for analysis and prediction purposes, mainly in the economics and

© Springer International Publishing AG, part of Springer Nature 2018
P. M. Buscema et al., *Artificial Adaptive Systems Using Auto Contractive Maps*, Studies in Systems, Decision and Control 131, https://doi.org/10.1007/978-3-319-75049-1_7

finance fields [13–17]. Being a weighted undirected graph, the MST provides a sort of narrative built into the data structure, which translates the complexity of the many-to-many relationships among variables of the database into sequences of transitions ('narratives') along the graph. In particular, the MST represents a minimal energy association among the variables, that is, the lowest informational cost associated to the graph-theoretical representation of the dataset's complexity. We also exploit the MST organizational properties in a recursive way, via a second fitness criterion that generates sequences of MSTs through pruning of the information at each layer. Algorithms perform differently with respect to the two fitness criteria. Here the two fitness are the MST graph and the consequence of disconnecting the various parts of the graph. This provides us with more insights as to their appropriateness for various types of datasets. A new method to evaluate the MST fitness of each algorithm will be presented [18]. For a comprehensive review of data mining and deep architectures in machine learning see [19–25].

7.2   Various Methods that We Compare

This section introduces the set of algorithms compared using our MST Fitness approach if each classical method is converted to weights. How this is done is presented next as we introduce the methods whose likeness and difference will be analysed.

A. **Principal Component Analysis (PCA)**: PCA is a classic multivariate linear analysis approach [4, 5]. In order to extract the MST from PCA, we compute the linear correlation of pairs of patterns defined by the PCA scores of their components according to the scheme outlined in Fig. 7.1.

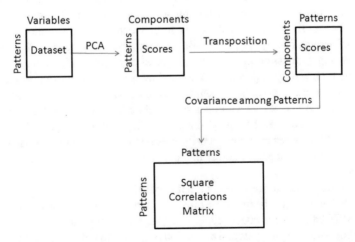

Fig. 7.1 Determination of the square correlations matrix for PCA

The calculation of the correlations is:

$$W_{i,j}^{[PCA]} = g\,(PCA\_Score)\,; \tag{7.1}$$

where:

$W_{i,j}^{[PCA]}$ = covariance Coefficient among records;
$-1 \le W_{i,j}^{[L]} \le 1; \quad i, j \in [1, 2, \ldots, M]$
$M =$ Number of patterns.

B. **Linear Correlation (LC)**: The LC is a classic Pearson covariance among patterns

$$W_{i,j}^{[LC]} = \frac{\sum_{k=1}^{N} \left(x_{i,k} - \bar{x}_i\right) \cdot \left(x_{j,k} - \bar{x}_j\right)}{\sqrt{\sum_{k=1}^{N} \left(x_{i,k} - \bar{x}_i\right)^2 \cdot \sum_{k=1}^{N} \left(x_{j,k} - \bar{x}_j\right)^2}}, \tag{7.2}$$

where:

$-1 \le W_{i,j}^{[LC]} \le 1; \quad i, j \in [1, 2, \ldots, M]$
$M =$ Number of patterns.

C. **Prior Probability Algorithm (PPA)**: The PPA a well-known algorithm that measures the frequency of concordances and discordances among patterns [26, 27] according to:

$$W_{i,j}^{[PPA]} = -\ln \left(\frac{\sum_{k=1}^{N} x_{i,k} \cdot \left(1 - x_{j,k}\right) \cdot \cdot \sum_{k=1}^{N} \left(1 - x_{i,k}\right) \cdot x_{j,k}}{\sum_{k=1}^{N} x_{i,k} \cdot x_{j,k} \cdot \sum_{k=1}^{N} \left(1 - x_{i,k}\right) \cdot \left(1 - x_{j,k}\right)} \right); \tag{7.3}$$

where:

$$-\infty \le W_{i,j}^{[PPA]} \le +\infty; \quad x \in [0, 1] \quad i, j \in [1, 2, \ldots, M].$$

D. **Euclidean Similarity (EU) and Manhattan Similarity (MA)**: two standard ways to implement Minkowski distance and connections weights given by:

$$W_{i,j}^{[MK^q]} = 1 - \left(\sum_{k=1}^{N} \left|x_{i,k} - x_{j,k}\right|^q \right)^{\frac{1}{q}}; \tag{7.4}$$

where:

$$0 \le W_{i,j}^{[MK^q]} \le 1; \quad x \in [0, 1]; \quad i, j \in [1, 2, \ldots, M].$$

$$\begin{cases} q = 1 \, Manhattan \\ q = 2 \, Euclidean \end{cases}$$

E. **Jaccard Similarity (JA)**: The JA method is another well-known equation that measures the intersection between sets of features [28, 29]:

$$W_{i,j}^{[J]} = \frac{\sum_{k=1}^{N} x_{i,k} \cdot x_{j,k}}{\sum_{k=1}^{N} x_{i,k} \cdot (1 - x_{j,k}) + \sum_{k=1}^{N} (1 - x_{i,k}) \cdot x_{j,k} + \sum_{k=1}^{N} x_{i,k} \cdot x_{j,k}};$$

(7.5)

where:

$$0 \le W_{i,j}^{[J]} \le 1; \quad x \in [0, 1]; \quad i, j \in [1, 2, \dots, M].$$

F. **Auto Contractive Map (Auto-CM)**: The Auto-CM method found in Chap. 3 computes nonlinear associations and similarities among patterns and variables according to specific contractive equations [10, 11]. Auto-CM provides a bottom-up approximation of the matrix of tensors among the variables or among the patterns in the dataset [11, 30]. Auto-CM has been applied with success in many fields [31–45] and its output goes directly into MST (see Chap. 4).

$$W_{i,j}^{[AutoCM]} = f^{[AutoCM]} (x_i, x_j);$$

(7.6)

where:

$$0 \le W_{i,j}^{[AutoCM]} \le 1; \quad x \in [0, 1]; \quad i, j \in [1, 2, \dots, M].$$

G. **Self-Organizing Map (SOM)**: The SOM is a very well-known Artificial Neural Network that projects onto a 2 dimensional matrix the similarity among patterns [9]. In order to extract the MST among patterns, we first transpose the dataset matrix and then we train the SOM using patterns as variables. Eventually, we extract the relations among patterns by means of the following equations.

M	Number of patterns of dataset, $i, j \in M$;
Q	Number of SOM cells (Rows × Columns), $k \in Q$;
$c_{i,j}$	i-th variable in the k-th cell;
c_T	Set of the non-empty cells;
$d_{i,j}$	Euclidean distance among patterns;
D^*	Max distance among patterns;
$f(x)$	A linear scaling function $-1 \le x \le +1$;
$W_{i,j}^{[SOM]}$	New SOM weights matrix

$$d_{i,j} = \sqrt{\sum_{k}^{c_k = C_T} (c_{k,i} - c_{k,j})^2};$$

$$D^* = Max_{i,j} \{d_{i,j}\}$$

$$W_{i,j}^{[SOM]} = f (D^* - d_{i,j})$$

(7.7)

Our test roster then consists of eight unsupervised algorithms:

a. Principal Component Analysis (PCA);
b. Linear Correlation (LC);
c. Prior Probability Algorithm (PPA);
d. Euclidean Similarity (EU);
e. Manhattan Similarity (MA);
f. Jaccard Similarity (JA);
g. Auto Contractive Map (Auto-CM);
h. Self Organizing Map (SOM).

7.3   The Datasets Used in the Comparative Evaluations

Having introduced the different algorithms that we want to test, we now come to the datasets we are using for the evaluation. Most of them are well-known artificial datasets, commonly employed as benchmarks, plus some more ones to extend the test to real-world problems and data.

A. Alphabet: is a small dataset of 36 alphabet signs, drawn into a 7 × 7 box (see Fig. 7.2). In our test, the algorithms will read the 36 patterns, and the corresponding 49 cells (7 × 7) as coordinates of each pattern. Thus Alphabet is a 36 × 49 dataset.

B. Foods: reports yearly consumption levels of nine types of foods in sixteen European countries in a specific year (see Fig. 7.3). The algorithms will read the 16 countries (patterns), using the 9 types of foods as coordinates for each country. Thus, Foods is a 16 × 9 dataset.

C. Gangs: is a famous dataset, widely used after its introduction in early key references in the field [46, 47]. It describes 14 attributes of the 27 members of the two gangs of West Side Story (Jets and Sharks, see Fig. 7.4). The algorithms will read as patterns the gang members, and their attributes as coordinates. Thus, Gangs is a 27 × 14 dataset.

D. Donald Duck: is an artificial dataset of seven famous Disney characters appearing in eight pictures (see Fig. 7.5). The algorithms will read the eight pictures as patterns, and the presence/absence of characters as coordinates. Thus, Donald Duck is an 8 × 7 dataset.

E. Rooms is a dataset featuring the a priori probability with which 50 different objects/attributes fit into 6 types of different rooms (Fig. 7.6). The algorithms will see the 50 objects/attributes as patterns, and the types of rooms as their coordinates. Thus, Rooms is a 50 × 6 dataset.

F. Italian Cities: is a dataset that describes 95 Italian Province capitals in the year 2000 by means of 43 socio-economic indicators (see Fig. 7.7). The algorithms will read the 95 Italian cities as patterns, and the 43 indicators as coordinates. Thus, Italian Cities is a 95 × 43 dataset.

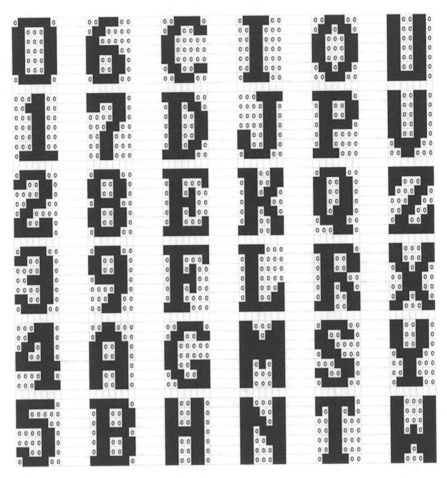

Fig. 7.2   The Alphabet dataset. 36 digits in a 7 × 7 box

We have therefore six datasets in total, with different characteristics. Alphabet, Gang, Donald Duck, and Rooms are a heterogeneous group of artificial datasets. Alphabet is about character recognition and visual pattern classification. Gangs is about deterministic attributes for two mutually exclusive classes. Donald Duck is about association rules for a group of objects and Rooms is about probabilistic attributes for six partially overlapping classes. All of them present well-known difficulties for unsupervised algorithms, as the patterns have been suitably designed so as to make clustering and dataset analysis challenging. An algorithm that would perform uniformly well for all of them would therefore emerge as a robust, flexible and powerful tool. The two datasets related to real-world phenomena, namely Foods and Italian Cities, are interesting in that they reflect complex associations of causal factors, but also because they refer to well-known and studied geo-social entities such as European countries and Italian Province capitals. The Foods dataset is based upon macro data at the national level, whereas Italian Cities contains data at a much

	Cereals	Rice	Potatoes	Sugar	Vegetables	Meat	Milk	Butter	Eggs
Belgium	72.2	4.2	98.8	40.4	103.2	102	80	7.7	14.2
Denmark	70.5	2.2	57	39.5	50	105.8	145.2	4.1	14.3
Germany	71.3	2.3	74.1	37.1	83.1	97.2	90.7	6.9	14.8
Greece	109.8	5.4	90	30	229.5	77.1	63.1	0.9	11.3
Spain	71.4	5.8	107.8	26.8	191.7	102.1	98.4	0.6	15.3
France	73	4.3	78.2	34.1	95	110.5	98.9	8.9	15
Ireland	93.4	3.2	151.5	34.8	55	105	185.9	3.4	11.4
Italy	110.2	4.8	38.6	27.9	181.9	88	65	2.4	11.1
Netherland	54.6	5	86.7	39.7	99	89.4	136.2	5.4	10.7
Portugal	86	5.7	106.6	29.4	100	75.5	96	1.5	7.7
Gr.Britain	74.3	4.5	94.1	39.8	60	74.4	129.3	3.2	10.8
Austria	68.7	4.2	62.6	37.1	81.9	93.4	121.3	4.3	13.4
Finnland	70.1	5.4	61.6	35.7	52.6	65	208.4	5.8	10.9
Island	79.7	1.9	50.2	54.9	50	71.7	205.6	4.6	11.3
Norway	76.9	3.5	73.2	37.3	48.3	54.9	176.5	2.1	11.3
Sweden	69.3	4.3	70	37.5	48.5	60.5	154.1	5.7	12.9

Fig. 7.3 The Foods dataset: yearly consumption levels of 9 types of foods (tons) in 16 European countries for a given year

smaller territorial scale. They feature therefore different levels of aggregation from the socio-spatial point of view, adding a further element of heterogeneity.

7.4 Evaluation Methods

We presented various methods to analyze different datasets. In this section we want to introduce the method we used to compare them. Our approach is based on the computation of fitness values that, starting from the MST connections, can measure the amount of similarity hidden in the dataset and detected by each method.

As we said in Chap. 2, there isn't a *gold standard* to compare with the results of an unsupervised method so the evaluation task becomes crucial to understand the goodness of the outputs.

Our MST Fitness approach [18] is based upon two indices, measuring two complementary fitness aspects:

a. *Main MST Fitness*: an index measuring the amount of well weighted similarities coded by the algorithm into the corresponding MST;
b. *Recursive MST Fitness*: an index measuring the amount of well weighted similarities coded by the algorithm in the recursively derived MSTs according to the procedure specified below.

In order to attain those fitness values some quantities are needed. At first all the dataset values must be linearly scaled into the [0, 1] range. So, we need to carry out the following transformation:

$$Ds = \{x_i\}_{i=1}^{M} \rightarrow D = \{v_i\}_{i=1}^{M} \, ;$$
$$\text{where: } v \in [0, 1] \, . \tag{7.8}$$

Name	Gang	Age	Education	Status	Occupations
ART	Jets	40	JuniorSchool	Single	Pusher
AL	Jets	30	JuniorSchool	Married	Burglar
SAM	Jets	20	College	Single	Bookie
CLYDE	Jets	40	JuniorSchool	Single	Bookie
MIKE	Jets	30	JuniorSchool	Single	Bookie
JIM	Jets	20	JuniorSchool	Divorced	Burglar
GREG	Jets	20	HighSchool	Married	Pusher
JOHN	Jets	20	JuniorSchool	Married	Burglar
DOUG	Jets	30	HighSchool	Single	Bookie
LANCE	Jets	20	JuniorSchool	Married	Burglar
GEORGE	Jets	20	JuniorSchool	Divorced	Burglar
PETE	Jets	20	HighSchool	Single	Bookie
FRED	Jets	20	HighSchool	Single	Pusher
GENE	Jets	20	College	Single	Pusher
RALPH	Jets	30	JuniorSchool	Single	Pusher
PHIL	Sharks	30	College	Married	Pusher
IKE	Sharks	30	JuniorSchool	Single	Bookie
NICK	Sharks	30	HighSchool	Single	Pusher
DON	Sharks	30	College	Married	Burglar
NED	Sharks	30	College	Married	Bookie
KARL	Sharks	40	HighSchool	Married	Bookie
KEN	Sharks	20	HighSchool	Single	Burglar
EARL	Sharks	40	HighSchool	Married	Burglar
RICK	Sharks	30	HighSchool	Divorced	Burglar
OL	Sharks	30	College	Married	Pusher
NEAL	Sharks	30	HighSchool	Single	Bookie
DAVE	Sharks	30	HighSchool	Divorced	Pusher

Fig. 7.4 The Gang dataset: 27 gang members. Each one defined by 14 attributes

where *Ds* is the original dataset, consisting of *N* variables and *M* records.

Then, the values of *Global similarity* and *Maximum similarity* are computed according to the following equations:

$$Sim_{i,j} = \sum_{k}^{M} \frac{v_{ik} \cdot v_{jk}}{\left[Max\left(v_{ik}, v_{jk}\right)\right]^2}; \quad if: \forall v \in D, v \in [0, 1] \qquad (7.9)$$

$$Sim_{i,j} = \sum_{k}^{M} v_{ik} \cdot v_{jk}; \quad if: \forall v \in D, v \in \{0, 1\} \qquad (7.10)$$

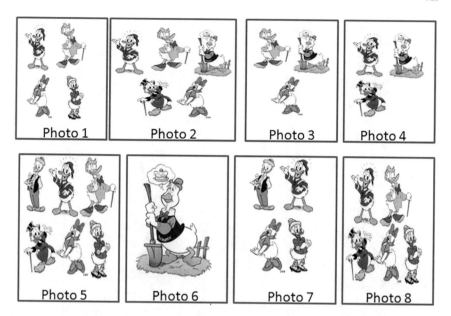

Fig. 7.5 The Donald Duck dataset: 7 Disney characters as appearing (or not) in 8 pictures

$$Sim\left(Global_D\right) = \sum_{i=1}^{N-1} \sum_{j=i+1}^{N} Sim_{i,j} \qquad (7.11)$$

$$Sim\left(Max\_MST\right) = \sum_{i=1}^{N-1} Rank_i^{Sup}\left(Sim_{i,j}\right) \qquad (7.12)$$

$Sim\left(Global\_D\right)$ measures the global similarity that can be detected in the entire dataset while $Sim(Max\_MST)$ calculates the quantity of similarity in the $N-1$ most correlated variables in the assigned dataset. We remark that those quantities are intrinsic of each analyzed the dataset and independent from the algorithm used.

At this point the weights matrix $W_{i,j}$ generated by the algorithm A to be evaluated is considered. We proceed to scale it linearly into the unit interval [0,1]:

$$w_{i,j} \rightarrow m_{i,j}, [m_{i,j}] = \mathbf{m}$$

where:

$m_{i,j} \in [0,1]$;
$i, j \in N$.

Then the relevant MST is computed, according to the following notation:

$$Mst(V, E) = f\left(A, D, m_{i,j}\right);$$

Rooms(50x6)	RESTROOM	KITCHEN	BEDROOM	DININGROOM	OFFICE	HALL
DESK	0	0	0	0	1	0
ASHTRAY	0.444	0.222	0.667	0.889	0.889	0.667
CHAIRS	0	0.889	0.222	0.889	1	0
EASY-CHAIR	0.222	0.889	0.222	0	0.111	0.444
BOOKS	0	0.444	0.444	0.889	0.778	0.778
BOOKSHELF	0	0.444	0.111	0.889	1	0.778
PENS	0	0.556	0.222	0.444	1	0.444
PAPERS	0	0.556	0.111	0.444	1	0.444
SOAP	1	1	0	0	0	0
PHONE	0.111	0.333	0.889	0.889	1	0.667
TV	0	0.444	0.778	0.889	0.556	0
RADIO	0.556	0.556	0.778	0.333	0.556	0
HI-FI	0	0	0	0.778	0.222	0
COMPUTER	0	0	0	0.111	0.889	0
DISKS	0	0	0	0.111	0.889	0
FLOOR-LAMP	0	0.222	1	0.222	1	0
PICTURES	0	0.222	0.667	1	0.556	1
CLOCK	0.667	0.889	1	1	1	0.778
SOFA	0	0	0	1	0.111	0
WINDOWS	0.556	1	1	1	1	0.333
BASKET	1	1	0	0	1	0.444
GADGETS	0.556	0.444	0.667	0.778	0.333	0.556
CEILING	1	1	1	1	1	1
DRAPES	0.222	0.778	1	0.889	0.667	0.444
STOVE	0	1	0	0	0	0.667
TOILET	1	0	0	0	0	0
WALLS	1	1	1	1	1	1
MEDIUM	0.444	0.889	0.889	0.333	0.667	0.667
BED	0	0	1	0	0	0
ARM-CHAIR	0	0	0	0.889	1	0
SHOWER	1	0	0	0	0	0
SINK	1	1	0	0	0	0
SCALE	0.778	0.889	0	0	0	0
DOOR	1	0.667	1	0.444	1	0.444
SMALL	0.889	0.667	0.111	0	0.111	0.444
TYPE-WRITER	0	0	0	0	1	0
COFFE-CUP	0	1	0	0	0	0
COFFEPOT	0	1	0	0	0	0
DRESSER	0.222	0	1	0	0	0.889
OVEN	0	0.889	0	0	0	0
VERY-SMALL	0.778	0.444	0	0	0.111	0.222
BOOKCASE	0	0	0.111	0.778	0.889	0.778
REFRIGERATOR	0	1	0	0	0	0
VERY-LARGE	0.222	0.111	0.444	0.889	0.889	0.222
CARPET	0	0	0.889	0.889	0.556	0.667
FIREPLACE	0	0	0	0.778	0	0
TOASTER	0	0.889	0	0	0	0
BATHTUB	0.778	0	0	0	0	0
CLOTHES-HANGER	1	0	0	0.444	0.667	1
DRAWERS	0.778	1	1	0.444	1	0.667

Fig. 7.6 Rooms dataset: the a priori probability with which fifty objects/attributes fit into each of five types of rooms

Population	vehicles_(except_center)
Density	murders
birth_rate	car_thefts
Mortality	thefts_in_the_apartments
Tumors	bank_robberies
Immigrations	scams
Separations	pickpocketing_
Income	artistic_and_cultural_associations
bank_deposits	spending_on_sports_events
Insurance_policies	spending_for_plays
Guesthouses	Spending_for_the_cinema
price_of_housing	spending_for_gyms
Protests	bookstores
bankrupt_companies	public_green
new_businesses	drinking_water_consumption
bad_checks	solid_waste
Placement	separate_collection
placement_(below_24_years)	fuel_consumption
waiting_for_new_pensions	unleaded
waiting_for_mail	domestic_consumption_of_electricity
pending_home_phone	public_transport
students_secondary_schools	

Fig. 7.7 The Italian Cities dataset: 43 socio-economic attributes of 93 Italian Province capitals

where:

D The assigned dataset scaled in the unit interval;
A The analized algorithm;
$m_{i,j}$ The parameters found by the algorithm;
i, j The indices of the parameters: $i, j \in N$;
V The nodes (variables) of the tree;
E The undirected links of the tree.

The *weighted meaningfulness* can now be easily calculated by weighing products of variables by means of the algorithm A coding as expressed by the MST:

$$Mst\_Sim\,(A, D) = \sum_{i=1}^{N-1} \sum_{j=i+1}^{N} \sum_{k}^{M} m_{i,j} \cdot \left(v_{i,k} \cdot v_{j,k} \right); \quad if \left(v_{i,k}, v_{j,k} \right) \in E \quad (7.13)$$

and accordingly, the *Main Fitness* of A is defined as:

$$Main\_Mst\_Fitness(A) = \frac{Mst\_Sim\,(A, D)}{Sim\,(Max\_MST)}. \quad (7.14)$$

We can now likewise define a second fitness criterion which works upon the recursive generation of all the MSTs obtained through a sequence of pruning, through the following steps:

a. Extract from the square and symmetric matrix of the algorithm parameters, **m**, the first-level MST;
b. Remove from the matrix parameters, **m**, all the connections present in the first-level MST;
c. Compute a new MST only making use of the remaining parameters;
d. If it is impossible to compute a new MST in this way, then stop the process. Otherwise, determine the new MST and subsequently extract and remove the second-level MST parameters from the matrix **m**, and go back to the step c.

The *Recursive Fitness* of A is given by:

$$Recursive\_MST\_Fitness\,(A,\,D)$$

$$= \frac{1}{Sim\,(Global\_D)} \sum_{n=1}^{NumMst} \frac{1}{n} \sum_{i=1}^{N-1} \sum_{j=i+1}^{N} \sum_{k=1}^{M} m_{n,i,j}\left(v_{i,k} \cdot v_{j,k}\right) \qquad (7.15)$$

$$if\,\left(v_{i,k},\,v_{j,k}\right) \in E^{Mst_n};$$

where:

NumMst   Maximum depth of MST recursion in the **m** matrix;
Mst$_n$ n-th level of MST recursion in the **m** matrix;
$\frac{1}{n}$ weighing factor: similarities become less meaningful as recursion goes deeper.

The Main Fitness of A measures the amount of similarity that A coded respect to the maximum similarity codifiable in an MST. The Recursive Fitness measures the amount of similarity coded by all the recursive MSTs obtained from the weights matrix of algorithm A respect to the global similarity presents in the original dataset.

We can summarize this section saying that the Main and the Recursive Fitness provide two complementary and quantitative ways to evaluate the results of an unsupervised algorithm.

7.5   Comparative Results

We now present the results of the comparative tests among the eight algorithms for each specific dataset. We also report the MST of the best and the worst performing algorithms for each dataset, so as to provide visual evidence of the different levels of articulation of the implicit narratives captured by differently performing algorithms in the context of the MST Fitness approach. We will then summarize all of the results in a final discussion.

7.5.1 The Alphabet Dataset

Table 7.1 shows that the Auto-CM MST provides the best representation of this dataset. PPA and PCA rank worst in terms of Main Fitness. Recursive Fitness closely agrees with Main Fitness here, so no major trade-off between effectiveness and representativeness occurs. It is interesting to notice that Auto-CM achieves top score in Recursive Fitness despite presenting the minimal depth of MST recursions among the eight algorithms. The meaningfulness in its recursive structure is high enough to best capture the deep structure of the dataset by a few iterations. It is possible to cluster the Main Fitness values of the eight algorithms into four performance classes:

  i. Top group: Auto-CM;
 ii. Top Middle group: Manhattan Similarity, Jaccard Similarity, Linear Correlation, and Euclidean Similarity;
iii. Bottom Middle group: SOM;
 iv. Bottom group: PCA and PPA.

   Figure 7.8a, b shows, respectively, the MSTs of Auto-CM and PPA. A visual comparison suffices to show how PPA and Auto-CM exhibit very different organizational properties. For instance, in PPA, for some characters such as U and S, the graph presents very high degrees, and such characters serve as hubs form any others (H, J, A, V, O, S for U; 5, 9, 3, 6, U for S), failing to deliver a clear criterion for structural analogies between characters. Nothing similar is found in the Auto-CM as measured by MST, where more fine-grained distinctions among characters are systematically drawn out. Likewise, one can check that PPA weighs its MST edges in a non-intuitive way, whereas the weights of Auto-CM seem more consistent: in particular, Auto-CM builds strong associations throughout its MST map, whereas

Table 7.1 Comparative performance for the Alphabet dataset

Alphabet 36 × 49	Main fitness	Recursive fitness	MST recursions	Rank
Auto-CM	0.948384	0.259583	13	1
MA	0.828317	0.183919	15	2
JA	0.806877	0.165737	13	3
LC	0.804889	0.181944	16	4
EU	0.781635	0.169633	15	5
SOM	0.77068	0.155272	14	6
PCA	0.516146	0.153507	16	7
PPA	0.449114	0.080609	14	8

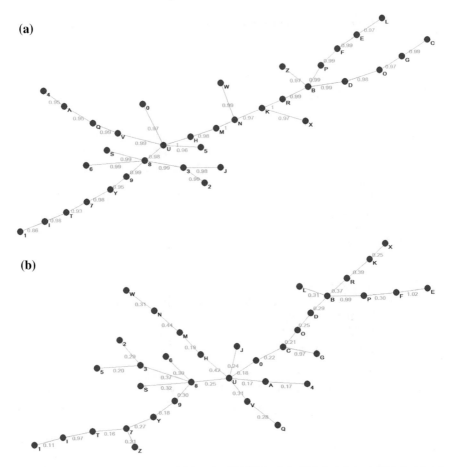

Fig. 7.8 **a** Alphabet dataset. Auto-CM first-level MST (top). **b** Alphabet dataset. PPA first-level MST (bottom)

PPA generally works through weak associations. That is, PPA's implicit narrative is precarious and likely to be structurally unstable.

7.5.2 The Foods Dataset

Results are shown in Table 7.2 and Auto-CM again takes first position, but this time PCA gets the second one, whereas PPA algorithm is the worst. Once again Main and Recursive Fitness scores are mutually consistent, although with some discrepancies. EU, for instance, does better than MA and LC on Main Fitness and vice versa for Recursive Fitness. Figure 7.9a, b report the MST measure of the best and worst algorithms. Notice how both algorithms cluster appropriately Mediterranean

Table 7.2 Comparative performance for the Foods dataset

Foods 16 × 9	Main fitness	Recursive fitness	MST recursions	Rank
Auto-CM	0.909562	0.281873	6	1
PCA	0.867567	0.240931	5	2
JA	0.786678	0.219442	7	3
EU	0.774002	0.18884	6	4
MA	0.753582	0.219946	7	5
LC	0.746917	0.196545	6	6
SOM	0.709001	0.194198	6	7
PPA	0.649643	0.172129	7	8

countries with similar food consumption habits (Spain, Italy, Greece and Portugal), whereas PPA does not recognize the subtler food style differences between two different group of Nordic countries (Denmark, Ireland and Iceland, on the one side, and Sweden, Finland and Norway, on the other side, namely, Western versus Eastern Nordic countries) in addition to placing the UK among the Nordic countries. Once again there are sharp differences in the MST weights. In particular the PPA weights are small enough to denote likely structural instability. As to performance classes with respect to Main Fitness, there are five groups: Auto-CM; then PCA; then JA, EU, MA, and LC; then SOM; and finally PPA.

7.5.3 The Gangs Dataset

Table 7.3 reports the results, which still place Auto-CM at the top, and PPA at the bottom. There is a small discrepancy between the two fitness measures, as SOM performs slightly better than Jaccard in Recursive Fitness and vice versa for Main Fitness, but the two measures are quite consistent overall. Here we have a particularly clear-cut performance criterion: the ability to correctly attribute Gang members. Auto-CM is the only algorithm that splits individuals into the two Gangs without any mistakes (Fig. 7.10a). PPA makes two mistakes (Fig. 7.10b), and once again the overall organization of the connections is more poorly organized, and through weaker links. In particular, the Ike-Mike link, which is the border between the two gangs in the Auto-CM narrative, is wrongly placed by PPA within the Jets field. PPA's border link instead ties next-closer opposite affiliates such as Doug and Neal. In terms of Main Fitness performance classes, we have: Auto-CM; PCA, LC, and MA; JA and SOM; EU; and finally PPA.

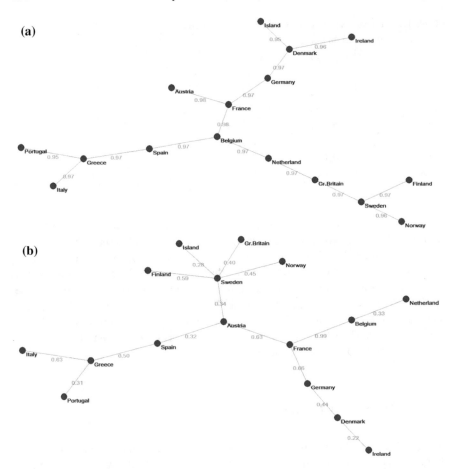

Fig. 7.9 a Foods dataset. First-level MST (top). **b** Foods dataset. PPA first-level MST (bottom)

7.5.4 The Donald Duck Dataset

Unlike the previous datasets, this dataset does not entail an obvious performance criterion. The issue is finding out sensible rules to organize the patterns. The best solution that emerges from the MST Fitness measure approach is once again Auto-CM (see Table 7.4). The Auto-CM as illustrated by MST organizes the pictures by optimizing two criteria simultaneously, the number of characters in each pictures and the presence/absence of the characters in all the pictures (see Fig. 7.11a). In particular, the role of Gus Goose is clearly singled out. He is the only one present in all the pictures with three or less characters, and in only one picture with more than three characters, which is correctly taken as the one linking the two main criteria. The worst solution is provided here by the Euclidean algorithm (see Fig. 7.11b), which somewhat mixes up the two rules, and collapses the structural complexity as

Table 7.3 Comparative performance for the Gangs dataset

Gangs 27 × 14	Main fitness	Recursive fitness	MST recursions	Rank
AutoCM	0.944177	0.204532	11	1
PCA	0.768907	0.119826	12	2
LC	0.768519	0.115954	11	3
MA	0.768519	0.115954	11	4
JA	0.655646	0.084139	11	5
SOM	0.643174	0.095327	11	6
EU	0.56886	0.073446	11	7
PPA	0.490611	0.085007	11	8

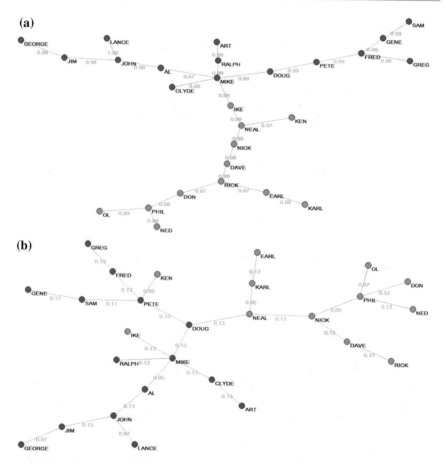

Fig. 7.10 a Gangs dataset. Auto-CM first-level MST (top). Red = Jets. Green = Sharks. **b** Gangs dataset. PPA first-level MST (bottom). Red = Jets. Green = Sharks

Table 7.4 Comparative performance for the Donald Duck dataset

Donald Duck 8 × 7	Main fitness	Recursive fitness	MST recursions	Rank
AutoCM	0.929778	0.314536	3	1
PPA	0.845608	0.233886	3	2
SOM	0.810645	0.218715	4	3
MA	0.797102	0.226708	3	4
JA	0.795777	0.211819	3	5
LC	0.777791	0.219381	3	6
EU	0.748141	0.201312	3	7
PCA	0.718155	0.212984	3	8

Table 7.5 Comparative performance for the Rooms dataset

Rooms 50 × 6	Main fitness	Recursive fitness	MST recursions	Rank
AutoCM	0.858457	0.202911	18	1
PCA	0.741176	0.137091	16	2
MA	0.703155	0.125097	20	3
LC	0.687179	0.126241	15	4
SOM	0.676335	0.108089	18	5
EU	0.656884	0.09946	18	6
JA	0.628829	0.090834	17	7
PPA	0.416506	0.052206	7	8

indicated by MST into a line. Here we find a major discrepancy between Main and Recursive Fitness, in that PCA scores second in terms of Recursive Fitness, but worst for Main Fitness. In terms of Main Fitness ranking, we have Auto-CM; PPA; SOM; MA and JA; then LC; EU; and finally PCA.

7.5.5 The Rooms Dataset

Also, for this dataset, Auto-CM is the best algorithm and PPA as the worst (see Table 7.5). The Auto-CM as measured by MST, shows a clear understanding of the structure of the dataset despite its intentional ambiguities. Objects are clustered according to the types of rooms in which are more likely to be found, and objects that typically can belong to different rooms are located between (see Fig. 7.12a for Auto-CM and Fig. 7.12b for PPA). For instance, in the case of PPA, the object 'desk' stands as an inefficiently dense hub which is directly linked to most of the stationery-related items, thereby missing out in the positioning of such objects in various possible rooms. For this dataset, moreover, Main and Recursive Fitness are perfectly concordant, and the performance classes are as follows: Auto-CM; PCA; MA, LC, and SOM; EU and JA; and finally PPA.

(a)

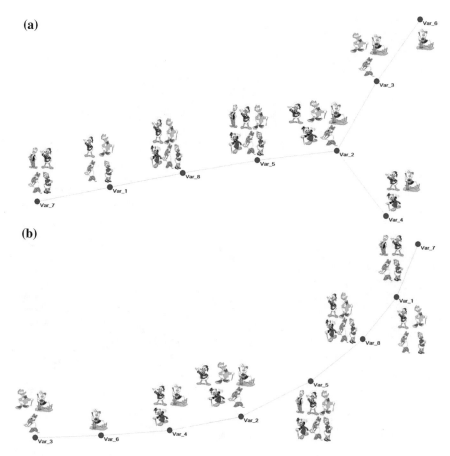

(b)

Fig. 7.11 **a** Donald Duck dataset. Auto-CM first-level MST (top). **b** Donald Duck dataset. EU first-level MST (bottom)

7.5.6 The Italian Cities Database

This is a particularly challenging dataset given the high number of coordinates for each pattern, which makes the classification space high dimensional. Auto-CM once more emerges as the best, whereas EU is the worst (see Table 7.6). Both the best and worst algorithms analyzed by MST correctly separate Northern cities from Southern ones (see Fig. 7.13a, b), but there are many fine-grained differences where Auto-CM is superior to EU. Whereas EU seems to classify mainly in terms of city population, with a tendency to group major cities together, Auto-CM makes sharp distinctions between cities of comparable size depending on their socio-economic characteristics. It is interesting for example how Auto-CM separates Verona from the other Veneto cities (Padova, Vicenza, Treviso, Venezia, Rovigo, with Padova as the hub), whereas EU groups them all together, with Vicenza as the hub. Whereas the other

Fig. 7.12 a Rooms dataset. Auto-CM first-level MST (top). **b** Rooms dataset. PPA first-level MST (bottom)

Veneto cities show a high level of territorial integration, with Padova as the natural hub both geographically and economically, Verona is an outlier in many respects, and gravitates more toward Eastern Lombardy than towards the rest of its own Region [48]. For this dataset, Main and Recursive Fitness yield slightly different results. Although the basic structure of the ranking is not affected. In terms of performance classes, we have Auto-CM; JA and PPA; LC and SOM; PCA and MA; and finally EU.

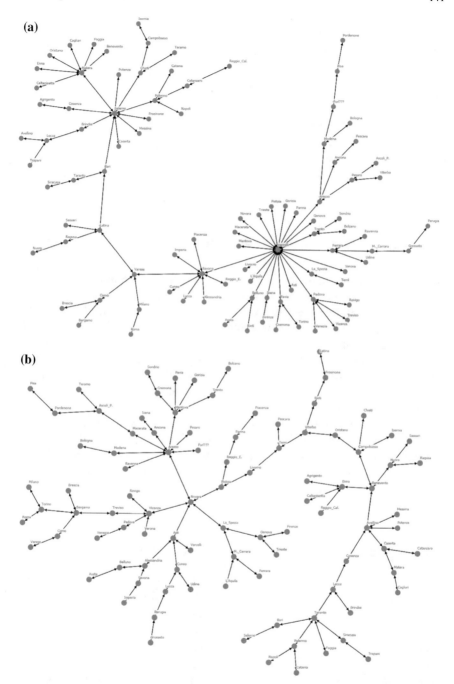

Fig. 7.13 a Italian Cities dataset. Auto-CM first-level MST (top). **b** Italian Cities dataset. EU first-level MST (bottom)

Table 7.6 Comparative performance for the Italian Cities dataset

Italian Cities 95 × 43	Main fitness	Recursive fitness	MST recursions	Rank
AutoCM	0.91327	0.273711	34	1
JA	0.809439	0.202933	32	2
PPA	0.802101	0.190297	32	3
LC	0.78398	0.217317	36	4
SOM	0.767341	0.203482	34	5
PCA	0.746912	0.170653	39	6
MA	0.736359	0.201756	34	7
EU	0.709693	0.188908	35	8

7.6 Final Discussion and Conclusions

We are now in the position to make a final evaluation and discussion of the results of our comparative test which we summarize in Table 7.7.

The main aspect that emerges is that Auto-CM is consistently the best algorithm for every dataset, uniformly sitting in a different performance class of its own. We can thus regard Auto-CM as a promising general purpose, unsupervised data mining tool. Moreover, in terms of our MST Fitness approach, Auto-CM generates subtly nuanced narratives with robust links between nodes. It remains to be proved whether Auto-CM is equally robust and performing on other types of datasets, but this is certainly a promising start.

PCA confirms its reputation as a reliable algorithm for clustering tasks (it is the second best for Foods, Gangs, and Rooms), but in some cases (Italian Cities, Alphabet, Donald Duck) its performance is very poor. One can thus regard it as a specialized algorithm, which should be used with care in poorly tested settings where its efficacy is not proven.

Table 7.7 Global ranking of the algorithms

Ranking	Alphabet	Foods	Gang	Donald Duck	Rooms	Italian Cities	Average	Variance
CM	1	1	1	1	1	1	1.0000	0.0000
PCA	7	2	2	8	2	6	4.0000	7.9000
JA	3	3	5	5	7	2	4.4000	3.3667
LC	4	6	3	6	4	4	4.6000	1.5000
MA	2	5	4	4	3	7	4.6000	2.9667
SOM	6	7	6	3	5	5	5.2000	1.8667
PPA	8	8	8	2	8	3	5.8000	8.1667
EU	5	4	7	7	6	8	6.4000	2.1667

PPA also shows a very high variability in performance, and fares very poorly when the number of patterns to classify is much bigger than the number of variables. It is the second best for Donald Duck and the third best for Italian Cities, but in all other cases it is the worst of the lot.

SOM yields a somewhat disappointing result here, contrary to expectations, and seems not very suitable for small dataset analysis. It is the third best for Donald Duck, but in all other cases its performance is mediocre.

MA and EU, despite their formal similarities, perform very differently. Apart from Foods and Italian Cities, MA consistently sits among the top four for all other datasets. EU, on the contrary, is merely the fourth best for Foods, and globally is the worst performing algorithm of the test. On the contrary, JA and LC perform in very similar ways in most datasets. JA is better in terms of global performance (its average rank is a bit higher), but LC is definitely more reliable in terms of rank variability. In terms of overall performance, we can thus define the following classes. The first is, unambiguously, Auto-CM (three full average ranks over the second best). PCA sits in the second class, but with a huge variability, the second higher in the whole sample. Then we have JA, LC, and MA. JA performs better but is more variable, whereas LC seems to deliver an optimal trade-off between performance and reliability in this class. Then, all in different classes, we have SOM; followed by PPA; and finally EU—although PPA has the largest variability of the whole sample whereas EU is relatively stable.

It is interesting to point out, in closing this section, how, with a few small exceptions, Main and Recursive Fitness tend to deliver mutually consistent results. On the other hand, it is not redundant to consider both criteria, in that they measure different aspects, and this could prove useful in further tests with a larger set of algorithms, especially if there is no clear top performer like Auto-CM. In principle, algorithms could score differently for the two criteria, as it happened for PCA in the Donald Duck dataset. Studying such discrepancies could yield better insight into the reasons behind the differential performance levels of the algorithms.

References

1. Strasser, B.J. 2012. Data Driven Sciences: From Wonder Cabinets to Electronic Databases. *Studies in History and Philosophy of Biological and Biomedical Sciences* 43: 85–87.
2. Wolfe, P.J. 2013. Making Sense of Big Data. *Proceedings of National Academy of Sciences* 110: 18031–18032.
3. Aflalo, Y., and R. Kimmel. 2013. Spectral Multidimensional Scaling. *Proceedings of National Academy of Sciences* 110: 18052–18057.
4. Bro, R., and A.K. Smilde. 2014. Principal Component Analysis. *Analytical Methods* 6: 2812–2831.
5. Abdi, H., and L.J. Williams. 2010. Principal Component Analysis. *Wiley Interdisciplinary Reviews: Computational Statistics* 2: 433–459.
6. Kuhnert, N., R. Jaiswal, P. Eravuchira, R.M. El-Abassy, B. von der Kammer, and A. Materny. 2011. Scope and Limitations of Principal Component Analysis of High Resolution of LC-TOF-

MS Data: The Analysis of the Chlorogenic Acid Fraction in Green Coffee Beans as a Case Study. *Analytical Methods* 3: 144–155.

7. Kalyagin, V.A., A.P. Koldanov, P.M. Pardalos, and V.A. Zamaraev. 2014. Measures of Uncertainty in Market Network Analysis. *Physica A* 413: 59–70.

8. Kalyagin, V.A., A.P. Koldanov, and P.M. Pardalos. 2014. A General Approach to Network Analysis of Statistical Data Sets. *Lecture in Computer Science*, vol. 8426, 88–97. Berlin: Springer.

9. Kohonen, T. 1995. *Self-organizing Maps*. Berlin: Springer.

10. Buscema, M. 2007. *Squashing Theory and Contractive Map Network*. Semeion Technical Paper #32, Rome.

11. Buscema, M., P.L. Sacco. 2010. Auto-contractive Maps, the H Function, and the Maximally Regular Graph (MRG): A New Methodology for Data Mining. In *Applications of Mathematics in Models. Artificial Neural Networks and Arts*. Chapter 11, ed. V. Capecchi et al. Berlin: Springer Science + Business Media B.V.https://doi.org/10.1007/978-90-481-8581-8_11.

12. Cormen, T.H., C.E. Leiserson, R.L. Rivest, and C. Stein. 2009. *Introduction to Algorithms*, 3rd ed. Cambridge MA: MIT Press.

13. Coelho, R., C.G. Gilmore, B. Lucey, P. Richmond, and S. Hutzler. 2007. The Evolution of Interdependence in World Equity Markets. Evidence from Minimum Spanning Trees. *Physica A* 376: 455–466.

14. Jang, W., J. Lee, and W. Chang. 2011. Currency Crises and the Evolution of Foreign Exchange Market: Evidence From Minimum Spanning Tree. *Physica A* 390: 707–718.

15. Zhang, Y., G.H.T. Lee, J.C. Wong, J.L. Kok, M. Prusty, and S.A. Cheong. 2011. Will the US Economy Recover in 2010? A Minimal Spanning Tree Study. *Physica A* 390: 2020–2050.

16. Dias, J. 2013. Spanning Trees and the Eurozone Crisis. *Physica A* 302: 5974–5984.

17. Sensoy, A., and B.M. Tabak. 2014. Dynamic Spanning Trees in Stock Market Networks: The Case of Asia-Pacific. *Physica A* 414: 387–402.

18. Buscema, M., and P. Luigi Sacco. 2016. MST Fitness Index and Implicit Data Narratives: A Comparative Test on Alternative Unsupervised Algorithms. *Physica A* 461: 726–746.

19. Tastle, W.J. (ed.). 2013. *Data Mining Applications Using Artificial Adaptive Systems*. New York: Springer Science + Business Media. https://doi.org/10.1007/978-1-4614-4223-3_1.

20. Buscema, M., and W. Tastle. 2015. An Innovative Theory of Analysis: Artificial Neural Network What-If Theory. *International Journal of Information Systems and Social Change* 6: 52–81.

21. Hinton, G.E., S. Osindero, and Y.-W. Teh. 2006. A Fast Learning Algorithm for Deep Belief Nets. *Neural Computation* 18: 1527–1554.

22. Bengio, J. 2009. Learning Deep Architecture for AI. *Machine Learning* 2: 1–127.

23. Maulik, U., and S. Bandyopadhyay. 2002. Performance Evaluation of Some Clustering Algorithms and Validity Indices. *IEEE Transactions on Pattern Analysis and Machine Intelligence* 24: 1650–1654.

24. Duda, T., and M. Canty. 2002. Unsupervised Classification of Satellite Imagery: Choosing a Good Algorithm. *International Journal of Remote Sensing* 23: 2193–2212.

25. Giraudel, J.L., and S. Lek. 2001. A Comparison of Self-organizing Map Algorithm and Some Conventional Statistical Methods for Ecological Community Ordination. *Ecological Modelling* 146: 329–339.

26. Arsuaga Uriarte, E., and F. Diaz Martin. 2005. Topology Preservation in SOM. *International Journal of Applied Mathematics and Computer Science* 1: 19–22.

27. Buscema, M. 1998. Constraint Satisfaction Neural Networks. *Substance Use & Misuse* 33: 389–408.

28. Consonni, V., and R. Todeschini. 2012. New Similarity Coefficients for Binary Data. *MATCH Communications in Mathematical and in Computer Chemistry* 68: 581–592.

29. Jaccard, P. 1901. Étude comparative de la distribution florale dans une portion des Alpes et des Jura. *Bulletin de la Société Vaudoise des Sciences Naturelles* 37: 547–579.

30. Jaccard, P. 1912. The Distribution of the Flora in the Alpine Zone. *New Phytologist* 11: 37–50. https://doi.org/10.1111/j.1469-8137.1912.tb05611.x.

31. Buscema, M., V. Consonni, D. Ballabio, A. Mauri, G. Massini, M. Breda, and R. Todeschini. 2014. K-CM: A New Artificial Neural Network. Application to Supervised Pattern Recognition. *Chemometrics and Intelligent Laboratory Systems* 138: 110–119.

32. Buscema, M. 2007. A Novel Adapting Mapping Method for Emergent Properties Discovery in Data Bases: Experience in Medical Field. In *2007 IEEE International Conference on Systems, Man and Cybernetics (SMC 2007)*. Montreal, Canada, October 7–10, 2007.

33. Buscema, M., and E. Grossi. 2008. The Semantic Connectivity Map: An Adapting Self-organizing Knowledge Discovery Method in Data Bases. Experience in Gastro-oesophageal Reflux Disease. *International Journal of Data Mining and Bioinformatics* 2 (4): 362–404.

34. Buscema, M., E. Grossi, D. Snowdon, and P. Antuono. 2008. Auto-Contractive Maps: An Artificial Adaptive System for Data Mining. An Application to Alzheimer Disease. *Current Alzheimer Research* 5: 481–498.

35. Buscema, M., C. Helgason, and E. Grossi. 2008. Auto Contractive Maps. H Function and Maximally Regular Graph: Theory and Applications. Special Session on Artificial Adaptive Systems in Medicine: Applications in the Real World. In *NAFIPS 2008 (IEEE)*, New York. May 19–22, 2008.

36. Licastro, F., E. Porcellini, M. Chiappelli, P. Forti, M. Buscema, et al. 2010. Multivariable Network Associated with Cognitive Decline and Dementia. *Neurobiology of Aging* 1: 257–269.

37. Grossi, E., G. TavanoBlessi, P.L. Sacco, and M. Buscema. 2012. The Interaction Between Cultures. Health and Psychological Well-Being: Data Mining from the Italian Culture and Well-Being Project. *Journal of Happiness Studies* 13: 129–148.

38. Licastro, F., E. Porcellini, P. Forti, M. Buscema, I. Carbone, G. Ravaglia, and E. Grossi. 2010. Multi factorial Interactions in the Pathogenesis Pathway of Alzheimer's Disease: A New Risk Charts for Prevention of Dementia. *Immunity & Ageing* 7: S4.

39. Buscema, M., F. Newman, E. Grossi, and W. Tastle. 2010. Application of Adaptive Systems Methodology to Radiotherapy. In *NAFIPS 2010*, July 12–14. Toronto, Canada.

40. Eller-Vainicher, C., V.V. Zhukouskaya, Y.V. Tolkachev, S.S. Koritko, E. Cairoli, E. Grossi, P. Beck-Peccoz, I. Chiodini, and A.P. Shepelkevich. 2011. Low Bone Mineral Density and Its Predictors in Type 1 Diabetic Patients Evaluated by the Classic Statistics and Artificial Neural Network Analysis. *Diabetes Care* 34: 2186–2191.

41. Gomiero, T., L. Croce, E. Grossi, L. De Vreese, M. Buscema, U. Mantesso, and E. De Bastiani. 2011. A Short Version of SIS (Support Intensity Scale): The Utility of the Application of Artificial Adaptive Systems. *US-China Education Review A* 2: 196–207.

42. Buscema, M., S. Penco, and E. Grossi. 2012. A Novel Mathematical Approach to Define the Genes/SNPs Conferring Risk or Protection in Sporadic Amyotrophic Lateral Sclerosis Based on Auto Contractive Map Neural Networks and Graph Theory. *Neurology Research International* (Art. ID 478560). https://doi.org/10.1155/2012/478560.

43. Grossi, E., A. Compare, and M. Buscema. 2014. The Concept of Individual Semantic Maps in Clinical Psychology: A Feasibility Study on a New Paradigm. *Quality & Quantity* 48: 15–35.

44. Coppedè, F., E. Grossi, M. Buscema, and L. Migliore. 2013. Application of Artificial Neural Networks to Investigate One-Carbon Metabolism in Alzheimer's Disease and Healthy Matched Individuals. *PlosOne*. https://doi.org/10.1371/journal.pone.0074012.

45. Street, M.E., M. Buscema, A. Smerieri, L. Montanini, and E. Grossi. 2013. Artificial Neural Networks and Evolutionary Algorithms as a Systems Biology Approach to a Data-Base on Fetal Growth Restriction. *Progress in Biophysics and Molecular Biology* 113: 433–438.

46. Compare, A., E. Grossi, M. Buscema, C. Zarbo, X. Mao, F. Faletra, E. Pasotti, T. Moccetti, P.M.C. Mommersteeg, and A. Auricchio. 2013. Combining Personality Traits with Traditional Risk Factors for Coronary Stenosis: An Artificial Neural Networks Solution in Patients with Computed Tomography Detected Coronary Artery Disease. *Cardiovascular Psychiatry and Neurology* (Art. ID 814967). http://dx.doi.org/10.1155/2013/814967.

47. McClelland, J.L. 1981. Retrieving General and Specific Information from Stored Knowledge of Specifics. *Proceedings of the Third Annual Meeting of the Cognitive Science Society* 170–172.

48. McClelland, J.L., D.E. Rumelhart, and G.E. Hinton. 1986. The Appeal of Parallel and Distributed Processing. In *Parallel and Distributed Processing*, vol. 1, ed. D.E. Rumelhart, and J.L. McClelland, 3–45. Cambridge, MA: MIT Press.
49. Adobati, F., A. Azzini, F.C. Pavesi. 2011. Il 'triangolo' Milano-Bergamo-Brescia: verso nuovi equilibri sub-regionali. XXXII Conferenza Italiana di Scienze Regionali, Turin.

Chapter 8
Auto-CM as a Dynamic Associative Memory

Abstract We look at how to use Auto-CM in the context of datasets that are changing in time. We modify our approach while keeping the original philosophy of Auto-CM.

8.1 Introduction

This section begins by adding an excitatory and inhibitory capability into Auto-CM. This allows Auto-CM to function as an auto-associative memory as will be demonstrated. A time dimension can be added to the auto-associative memory to make it a dynamic auto-associative memory. Here, by "dynamic", we mean the capability to deal with databases that are changing rather than an explicit dependence on time although the way changes occur may be time based. After the learning phase Auto-CM ANN has projected all its knowledge about the training dataset into its weight matrix. Different from the other Auto Associative ANNs, when Auto-CM terminates its learning phase, all its output nodes tend to zero as a side effect of learning itself. This behaviour is biologically plausible. It is useless to repeat mechanically the input vector in the output layer, once the artificial synapses (weights matrices) are already perfectly tuned. The other ANNs, instead, (deep learning included), perform exactly in this autistic way. Auto-CM appears to understand when it has extracted from the data all the information it can detect. Consequently, Auto-CM shows no reaction in front of an already known input. And this behaviour is biologically reasonable. In fact, "intelligence" is fundamentally the capability to predict the future and to react promptly when our prediction is not validated from the reality [1].

In this chapter we want to show how to extract from the Auto-CM weight matrix the information learned during the training phase.

We propose two techniques for this job:

a. An easy method, we call "Recall in One Shot" (ROS, for short);
b. A complex method, we call Spin Net.

© Springer International Publishing AG, part of Springer Nature 2018   147
P. M. Buscema et al., *Artificial Adaptive Systems Using Auto Contractive Maps*, Studies
in Systems, Decision and Control 131, https://doi.org/10.1007/978-3-319-75049-1_8

8.2   Recall in One Shot (ROS) [2]

ROS is a simple technique to extract key information from the weights matrix of Auto-CM, once this ANN has learned a dataset.

We denote by $l(\cdot)$ the linear function, whose domain is between 0 and 1, acting on the value of the function argument, moving it in that range. By *Input* we mean the (incomplete) input vector used to activate the Auto-CM trained weights $w^{[AutoCM]}$ and by *Output* we mean the vector describing how Auto-CM reacts to a specific input vector.

The following equations describe how Auto-CM completes a generic input vector, using its trained weights matrix. We remind the reader that the incomplete input vector has to input directly into the hidden layer of Auto-CM, in order to activate the second weights matrix, w. The first set of equations describes the flow of the signal from the Input vector to the Output Vector that Auto-CM builds.

$$Net_i = \sum_{j}^{N} Input_j \cdot w_{i,j}^{[AutoCM]}; \tag{8.1}$$

$$Output_i = l(Net_i). \tag{8.2}$$

$l(\cdot) =$ linear scaling in the interval $[0,1]$;
$l(Net_i) = Scale_i \cdot Net_i + Offset_i$;
$Scale_i = \frac{1}{Max(Net_i) - Min(Net_i)}$
$Offset_i = -\frac{Min(Net_i)}{Max(Net_i) - Min(Net_i)}$

The next set of equations describes how Auto-CM activates the records of the dataset, using the output values as weights of the records themselves.

$$Net_i = \sum_{j}^{N} (2 \cdot Output_j - 1) R_{i,j}; \tag{8.3}$$

$$a_i = \frac{e^{Net_i} - e^{Net_i}}{e^{Net_i} + e^{Net_i}} = \tanh(Net_i). \tag{8.4}$$

$$OutputRec_i = \beta.a_i \tag{8.5}$$

where

$N =$ Number of Components of the input vector;
$\beta = \frac{1}{N}$;
$R_{i,j} =$ Value of the i-th Record in the j-th variable in the dataset;
$Output_i =$ Output of the i-th variable after the recall process;
$a_i =$ i-th record activation
$OutputRec_i =$ Output of the i-th record after the recall process.

Table 8.1 Dataset of Jets and Sharks

Name	Gang	Age	Education	Status	Occupations
ART	Jets	40	Junior School	Single	Pusher
AL	Jets	30	Junior School	Married	Burglar
SAM	Jets	20	College	Single	Bookie
CLYDE	Jets	40	Junior School	Single	Bookie
MIKE	Jets	30	Junior School	Single	Bookie
JIM	Jets	20	Junior School	Divorced	Burglar
GREG	Jets	20	High School	Married	Pusher
JOHN	Jets	20	Junior School	Married	Burglar
DOUG	Jets	30	High School	Single	Bookie
LANCE	Jets	20	Junior School	Married	Burglar
GEORGE	Jets	20	Junior School	Divorced	Burglar
PETE	Jets	20	High School	Single	Bookie
ERED	Jets	20	High School	Single	Pusher
GENE	Jets	20	College	Single	Pusher
RALPH	Jets	30	Junior School	Single	Pusher
PHIL	Sharks	30	College	Married	Pusher
IKE	Sharks	30	Junior School	Single	Bookie
NICK	Sharks	30	High School	Single	Pusher
DON	Sharks	30	College	Married	Burglar
NED	Sharks	30	College	Married	Bookie
KARL	Sharks	40	High School	Married	Bookie
KEN	Sharks	20	High School	Single	Burglar
EARL	Sharks	40	High School	Married	Burglar
RICK	Sharks	30	High School	Divorced	Burglar
OL	Sharks	30	College	Married	Pusher
NEAL	Sharks	30	High School	Single	Bookie
DAVE	Sharks	30	High School	Divorced	Pusher

How ROS works: a simple example—Let us consider the small dataset named Gang that we have seen before, and let us consider also the distribution of the dataset members into the two gangs, Jets and Sharks (Tables 8.1 and 8.2).

We input to an ANN after Auto-CM has learned this dataset, the following vector (Fig. 8.1a).

This vector presents only one component activated (Jets = 1) and all the other variables with zero. In other words, we want to understand if a trained Auto-CM is able to complete an incomplete input pattern providing the ANN,

a.   The complete prototype variable of a Jets member;

Table 8.2 Statistics of the Jets and Sharks Dataset

Variable	Jet	Shark	Jet (%)	Shark (%)
20s	9	1	60.00	8.33
30s	4	9	26.67	75.00
40s	2	2	13.33	16.67
JH	9	1	60.00	8.33
HS	4	7	26.67	58.33
COL	2	4	13.33	33.33
Single	9	4	60.00	33.33
Married	4	6	26.67	50.00
Divorced	2	2	13.33	16.67
Pusher	5	4	33.33	33.33
Bookie	5	4	33.33	33.33
Burglar	5	4	33.33	33.33

(a)

Jet**	Sharks	20's	30's	40's	JH	COL	HS	Single	Married	Divorced	Pusher	Bookie	Burglar
1	0	0	0	0	0	0	0	0	0	0	0	0	0

(b)

Gang		Age			Education			Status			Profession		
Jet**	Sharks:	20's:	30's:	40's:	JH:	COL:	HS:	Single:	Married:	Divorced:	Pusher:	Bookie:	Burglar:
1	0	0.976504	0.778105	0.685589	0.975378	0.663404	0.787941	0.954431	0.815029	0.662244	0.881934	0.863017	0.84416

Fig. 8.1 a The incomplete input vector. **b** The Jets prototype generated by auto-CM

b. And which members of the dataset belong to this prototype and with which degree. Applying the first set of equations, the input vector is completed and displayed in Fig. 8.1b.

The Auto-CM answer is very reasonable. It defines the degree of membership of any Jets member to each variable, and it outlines also the most representative Jets features: 20's, JH, Single and Pusher (the highest value of each category in red). Now, using this Output and the second set of equations, Auto-CM shows us which members are more representative of the Jets gang (Fig. 8.2).

Figure 8.2 shows that the Jets members are more activated than the Sharks members, and two of the Sharks more activated (Ken and Ike) are the Sharks records most similar to the Jets prototype (Ken is the only Sharks to be 20's and Ike is the only Sharks to have the JH degree). We have similar results if we are looking for the prototype of the Sharks member (Figs. 8.3a, b and 8.4).

We would get reasonable results inputting a different incomplete vector to Auto-CM; ROS algorithm will always complete the input providing complete patterns showing the key information embedded into the dataset. Thus, the ROS algorithm is an easy and suitable technique to extract from a trained Auto-CM all the knowledge trapped in its weights matrix.

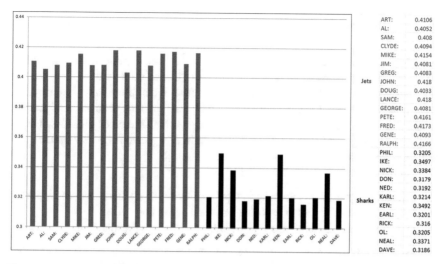

Fig. 8.2 Auto-CM records activation for Jets prototype

(a)

Jet	Sharks**	20's	30's	40's	JH	COL	HS	Single	Married	Divorced	Pusher	Bookie	Burglar
0	1	0	0	0	0	0	0	0	0	0	0	0	0

(b)

Gang		Age			Education			Status			Profession		
Jet**	Sharks	20's	30's	40's	JH	COL	HS	Single	Married	Divorced	Pusher	Bookie	Burglar
0	1	0.391958	0.96924	0.731081	0.413257	0.860295	0.954373	0.792114	0.935127	0.720173	0.844635	0.84778	0.879575

Fig. 8.3 **a** Incomplete input for the Sharks prototype. **b** The Sharks prototype generated by Auto-CM

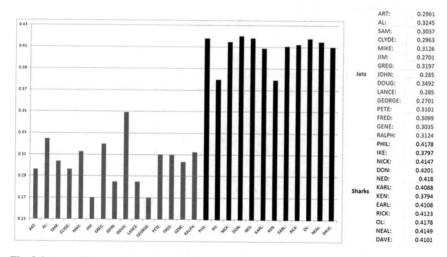

Fig. 8.4 Auto-CM records activation for Sharks prototype

8.3   Spin Net [3]

Spin Net is an adaptive algorithm that is able to extract the key information embedded in the trained Auto-CM weights matrix. Spin Net does this job working as a dynamic auto associative memory. In other words, when an incomplete Input is introduced into Spin Net system, the algorithm shows the path by which each variable tends to attain its stable value by dynamically interacting with all the other variables. The set of paths at the end of each performance of Spin Net generates a new dataset with different information from the original one.

Spin Net equations have three aims:

a. To calculate the state of each node after the incomplete pattern has been inputted;
b. To calculate the output of each node at each activation cycle;
c. Define an optimal number of cycles able to terminate the algorithm.

To these ends the following equations are applied. They are Eqs. (8.6), (8.7) which calculate the internal state of each node from the pattern inputted.
Where,

$p \in \{1, \ldots, M\}$;
$i, j \in \{1, 2 \ldots, N\}$;
N = Input Number of each pattern;
M = Pattern Number;
$n \in [0, +\infty)$; Index for cycles;
$w_{i,j}^{[T]} \in [0, 1]$; Trained AoutCM weight matrix;
$C = \sqrt{N}$;
$Input_i^{[n=0]}$ = External i-th Impulse at cycle $n = 0$;
$s_i^{[n]}$ = Internal State of the i-th node at n-th cycle;
$u_i^{[n]}$ = Activation of the i-th node at n-th cycle;
$Outut_i^{[n]}$ = Final Output of the i-th node at n-th cycle;
$\Delta^{[n]}$ = Total Energy difference between external input and internal state at n-th cycle;
$l(\cdot)$ = Linear Function scaling values into [0,1] interval.

$$Net_i^{[n]} = \frac{1}{N} \sum_{j}^{N} input_j^{[n]} \left(1 - \frac{w_{i,j}^{[T]}}{C} \right) \tag{8.6}$$

$$s_i^{[n]} = l \left(input_i^{[n]} \left(1 - \frac{Net_i^{[n]}}{C^2} \right) \right) \tag{8.7}$$

This is followed by Eq. (8.8) which defines the global difference between the input pattern and the internal states of the networks at each cycle.

$$\Delta^{[n]} = \sum_{i}^{N} \left(s_i^{[n]} - input_i^{[n]} \right)^2. \tag{8.8}$$

Equation (8.9) linearly stretches the internal state of each node, (that we have called the *activation value*) only when the original input pattern (at n = 0) is zero.

$$\begin{cases} u_i^{[n]} = l\left(s_i^{[n]} \right) ; input_i^{[n=0]} = 0 \\ u_i^{[n]} = s_i^{[n]}; \qquad input_i^{[n=0]} \neq 0. \end{cases} \tag{8.9}$$

Since dealing with zero values could be problematic, during the execution when $input_i^{[n=0]} = 0$ it is slightly modified into $input_i^{[n=0]} = 0.000001$.

Equation (8.9) means that another linear scaling function is applied only to the states corresponding to the units which $input_i^{[n=0]} \neq 0$. Equations (8.10) and (8.11) calculate the output value of each node of the network and set up the next recursive input with the value of the actual activation value,

$$Output_i^{[n]} = u_i^{[n]} \cdot \left(1 - \frac{\sum_j^N u_j^{[n]} \left(1 - \frac{w_{i,j}^{[T]}}{C} \right)}{C^2} \right) \tag{8.10}$$

$$Input_i^{[n+1]} = u_i^{[n]}. \tag{8.11}$$

Lastly, Eq. (8.12) controls the algorithm termination. When the energy of the actual input pattern is similar to the energy of the internal states of the networks, then the process is stopped. That is:

$$\text{while } \left(\Delta^n > \varepsilon \right) ; \text{repeat and set } n = n + 1. \tag{8.12}$$

Equations (8.8) and (8.12) represent a novelty in the landscape of artificial auto associative memories. Most known auto associative memories would reach their stable state updating the values of their units at each new cycle with a new delta value. In this way they increase or decrease their previous values. Constraints Satisfaction Networks (CS) or Interactive Activation and Competition (IAC) work in this way [4–12]. These kinds of ANNs, in fact, need of excitatory weights and inhibitory weights and the activation of each unit is defined by the accumulation of delta updating at each evolutionary cycle. The following equations represents the basic dynamics of Constraints Satisfaction Neural Networks (Eqs. 8.13–8.15), and Eq. (8.16) is the cost function to maximize.

$$Net_i(n) = \sum_{j}^{N} u_j(n).w_j + Bias_i + InputExt_i; \tag{8.13}$$

$$\begin{cases} if \ (Net_i(n) > 0) \ \Delta u_i(n) = Net_i(n). \ (1 - u_i(n)) \\ \quad\quad else \quad\quad\quad\quad \Delta u_i(n) = Net_i(n)\Delta u_i(n) \end{cases} \quad (8.14)$$

$$u_i(n+1) = u_i(n) + \Delta u_i \quad\quad (8.15)$$

$$G(n) = \sum_i \sum_j w_{i,j}.u_i(n) + \sum_i Bias_i - \sum_j InputExt_i.u_i(n) \quad (8.16)$$

Fundamentally CS ANN is an adaptive algorithm that aims to maximize the energy. If all its weights had positive values, the CS output would set all its units at value 1, despite the external input assumed as starting point. Indeed, CS ANNs and the other Auto Associative NNs are adaptive algorithms that in order to maximize the information of their weights behave as they were a human memory process. But human memory seems to work much more as a recurrent interpretation of its internal states and of their synapses, more that the solution of an optimization problem.

During the Spin Net process, instead, each unit changes its value in order to minimize the difference between the inside energy of the networks (values of the internal states) and the energy outside of the network (initial external inputs). In order to optimize this goal, Spin Net at each cycle uses all the values of its internal states as "external" input for the next cycle (see Eqs. 8.9 and 8.11). In other words, Spin Net dynamically creates its external input. From a theoretical stand point, Spin Net executes a deep learning strategy. At any cycle it gives a new representation (by means its internal states) to its representation (internal state) of the cycle before. That is, Spin Net continually updates its understandings and it does not need inhibitory weights to work.

We call "Re-Entry" the technique used by Spin Net to input recurrently into the ANN the internal states that it has generated one cycle before.

It should be noticed that the Spin Net output isn't involved in the "Re-Entry" technique: we only use the internal states. The reason we do that is the belief that the output is just an elaboration of the internal states which contain the non-filtered information. In brief, we use as external input what the Net thinks, not what it says.

How Spin Networks: a simple example—We will use the same Gang dataset with the same input patterns: once to define the Jets prototype (Fig. 8.1a) and another time to define the Sharks prototype (Fig. 8.3a). Let us consider the Spin Net behaviour when we input the incomplete pattern for Jets members (Fig. 8.1a). After 2762 cycles the algorithm converges. Figure 8.5a, b shows the final attractor from the variables point of view. Figure 8.6 shows the same situation from the records' (data set patterns) point of view.

It is evident that Spin Net detects the fundamental features of the Jets proto-type in a more precise form with respect to the simplistic ROS algorithm. But, the most important information generated by Spin Net is in the process through which the variables interact with each other. Figure 8.7 shows the dynamics of this content addressable memory: Input variable "Jets" is fixed at 1 at each cycle, and the other variables interact with each other, negotiating dynamically and in parallel their values.

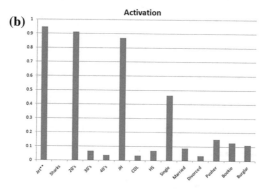

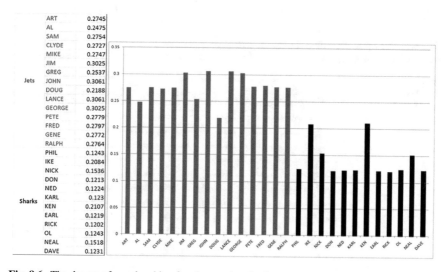

Fig. 8.5 **a** The Jet prototype generated by Spin Net. **b** Bar graph of the Jet prototype generated by Spin Net

Fig. 8.6 The degree of membership of each record to the Jet prototype generated by Spin Net

We can make the same experiment fixing at 1 the variable "Sharks" (according to the input vector showed in Fig. 8.3a). Figure 8.8a, b show the Spin Net answer after 2250 cycles from the variables point of view, and Fig. 8.9 shows the same attractor point from the patterns point of view. Finally, Fig. 8.10 shows the dynamics of variables negotiation during the Spin Net process. In this experimentation, as in the previous one, Spin Net works as a Content Addressable Memory. But also in this case the process and the results are meaningful.

Each time that we activate the interactive process regulated by Spin Net, a new dataset is generated. At each cycle, all the variables take a new value, according to

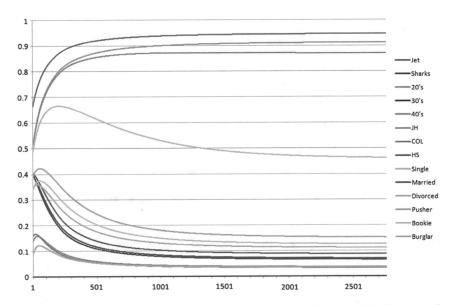

Fig. 8.7 The Spin Net output during the 2762 cycles through which the variables interact each other before to reach up a stable attractor. The input variable "Jets" is fixed at 1 at each cycle

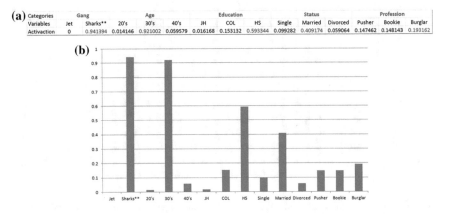

Fig. 8.8 a The Sharks prototype generated by Spin Net. **b** Bar graph of the Sharks prototype generated by Spin Net

the values they had at the cycle before and according to the trained Auto-CM weights matrix. This dataset presents new information in comparison to the original "Gang" dataset. This is because the dataset depends strongly on the variables that we have constrained with a specific value at the beginning of the process. Thus, the Spin Net process is the point of view of the effect of one variable on the other variables.

We generate a chain of states (output at each cycle) in the Gang dataset when we set to 1 the variable "Jets" before the process starts, where each state is an "effect" of

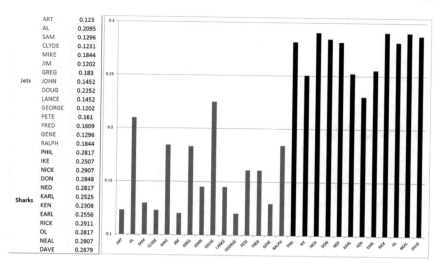

Fig. 8.9 The degree of membership of each record to the Sharks prototype generated by Spin Net

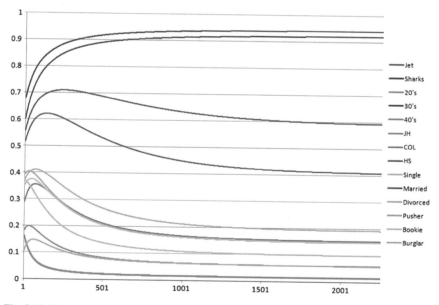

Fig. 8.10 The Spin Net output during the 2250 cycles through which the variables interact each other before to reach up a stable attractor. The input variable "Sharks" is fixed at 1 at each cycle

the previous one and it is also a "cause" of the next one. The same process happens when we set value of 1 to the variable "Sharks" at the beginning.

We have generated two sequential datasets in both cases, each one representing the artificial memory process from a specific point of view (Jets and Sharks viewpoints). At this point we can try to detect the specific rules governing these two specific processes of artificial memory, and, consequently, we can try to represent with two different directed and weighted graphs the point of view of Jets and the point of view of Sharks onto the same dataset. But to find out the specific rules generated by Spin Net, we need introduce a new algorithm we call Target Diffusion Mode (TDM).

8.4 Target Diffusion Mode (TDM) [13]

TDM equations can be applied to any dataset generated by any Auto Associative Memory Recall Process such as Spin Net. The main goal of TDM is to define the two cause-effect networks among the variables. More specifically:

• The direct graph of the cause-effect excitatory connections;
• The direct graph of the cause-effect inhibitory connections.

Let us imagine that $q_i^{[n]}$ is the value of the i-th variable at n cycle of the dataset generated by Spin Net. Let us also assume that the matrix, $\mathbf{q}$, is linearly scaled to $[0,1]$, that is $q \in [0,1]$. Now for each sequential pair of states of the dataset we define two quantities, α and β, in the following way (Eqs. 8.17 and 8.18):

$$\alpha_i^{[n,n+1]} = q_i^{[n+1]} - q_i^{[n]} \tag{8.17}$$
$$\beta_j^{[n,n+1]} = q_j^{[n+1]} - q_j^{[n]} \tag{8.18}$$

where

$i \neq j$

and

$i, j =$ two different variables of the dataset.

If N is the number of states of the dataset (that is the number of cycles necessary for Spin Net to converge), we can define a new quantity for any pair of variables (Eq. 8.19):

$$E_{i,j} = \sum_{n=1}^{N-1} \beta_j^{[n,n+1]} \quad if \left(\alpha_i^{[n,n+1]} > 0 \right) \tag{8.19}$$

Equation (8.19) calculates the strength that the i-th variable is the excitatory or inhibitory cause of the j-th variable. Equation (8.20) measures the possible correlation between any pair of variables in the whole dataset.

$$S_{i,j} = \sum_{n=1}^{N-1} q_i^{[n]} q_j^{[n+1]}. \tag{8.20}$$

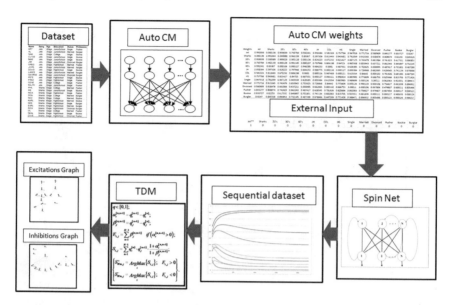

Fig. 8.11 From dataset to final graphs. Spin Net can generate each time a different dataset according to the external input

Consequently, Eq. (8.21) chooses the most likelihood excitatory cause and the most likelihood inhibitory cause for each variable.

$$\begin{cases} S^{+}_{W_{in,j}} = ArgMax_i \left\{ S_{i,j} \right\}; E_{i,j} > 0 \\ S^{-}_{W_{in,j}} = ArgMax_i \left\{ S_{i,j} \right\}; E_{i,j} < 0 \end{cases} \tag{8.21}$$

If we consider Eq. (8.21) we are able to generate two directed and weighted graphs. One is the network that has the most likely *excitatory* cause-effect relationship, and another one showing the networks of more likelihood *inhibitory* cause-effect relationship.

8.5   From Auto-CM to the Subjective Graphs

Figure 8.11 shows the entire process from a dataset processed by through Auto-CM followed by Spine Net to TDM graphs. This complex processing framework represents a new theoretical object, that is, an Artificial Adaptive System, whose components are different types of ANNs and Intelligent Algorithms.

We think that any intelligent data mining activity has to be based on Artificial Adaptive Systems because a single ANN or Learning Machine is not able to detect all the hidden information embedded in a dataset.

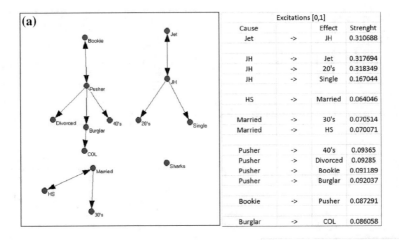

Excitations [0,1]			
Cause		Effect	Strenght
Jet	->	JH	0.310688
JH	->	Jet	0.317694
JH	->	20's	0.318349
JH	->	Single	0.167044
HS	->	Married	0.064046
Married	->	30's	0.070514
Married	->	HS	0.070071
Pusher	->	40's	0.09365
Pusher	->	Divorced	0.09285
Pusher	->	Bookie	0.091189
Pusher	->	Burglar	0.092037
Bookie	->	Pusher	0.087291
Burglar	->	COL	0.086058

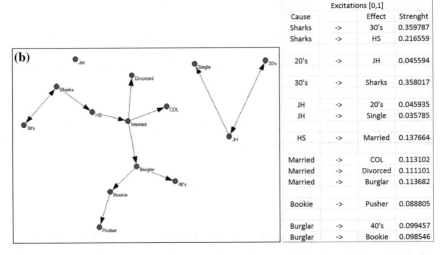

Excitations [0,1]			
Cause		Effect	Strenght
Sharks	->	30's	0.359787
Sharks	->	HS	0.216559
20's	->	JH	0.045594
30's	->	Sharks	0.358017
JH	->	20's	0.045935
JH	->	Single	0.035785
HS	->	Married	0.137664
Married	->	COL	0.113102
Married	->	Divorced	0.111101
Married	->	Burglar	0.113682
Bookie	->	Pusher	0.088805
Burglar	->	40's	0.099457
Burglar	->	Bookie	0.098546

Fig. 8.12 **a** Cause-effect excitatory graph from the "Jets" perspective. **b** Cause-effect excitatory graph from the "Sharks" perspective

Coming back to our Gang dataset and the two new datasets generated by Spin Net, we can project the graphs of the cause-effect excitations and the graphs of the cause-effect inhibitions from the point of view of Jets and Sharks. Figure 8.12a shows the excitatory graph among the variables from the point of view of "Jets".

Figure 8.12a shows three clusters and one isolated node. The node isolated is the variable "Sharks" because this variable is completely orthogonal to the "Jets" variable. The first cluster shows the most relevant variables from Jets point of view: the variables "Jets" and "JH" excite each other and, consequently, the variable "JH" excites the most representative features of the Jets prototypes which is 20's and Single.

The second cluster represents those variables which rarely constitute a "Jets" prototype, "Pusher" and "Bookie". The typical Jet is Divorced, 40's, Burglar and College.

The third cluster selects and organizes other three variables not specific for the Jets prototype: "Married" and "HS" that support each other, and the variable "30's", supported by the variable "Married".

Figure 8.12b shows the same situation from the "Sharks" point of view. The "Jets" variable appears with no connection to the others, as well as "Sharks" variable in Fig. 8.12a.

The first cluster of Fig. 8.12b shows the "genealogy" of the variables compatible with the "Sharks" prototype. The basic building blocks of this prototype are the variables "Sharks" and "30's"; then "Sharks" which generates "HS", which generates "Married", which generates "COL" and "Burglar"; then "Burglar" generates "40'" and "Bookie" and the last one generates "Pusher".

The second cluster, instead, shows the variables less typical for the "Sharks" profile, "20s", "JH" and "Single". It is interesting to observe that the "Jets" world from the point of view of the "Jets" gang is more specific than the "Sharks" world from the point of view of the "Sharks" gang. Figure 8.13a, b show the inhibitory graphs from "Jets" perspective and from "Sharks" perspective.

It is evident from these graphs that the inhibitory strategy of the variable "Sharks" is more modulated than the inhibitory strategy of the variable "Jets". However, from these inhibitory graphs a proportion might be extracted. In fact, if we mean $f(y) = y$ *from $f(.)$ point of view*, then we can write:

$$Jet(JH) : Sharks(HS) = Jets(Jets \rightarrow Single) : Sharks(Sharks \rightarrow Married).$$

What we have shown is that we have a method that is able to extract complex rules of weak signal modulations using a small dataset and only two simple external stimulations. This combination of artificial adaptive system (Auto-CM, Spin Net and TDM) seems to us more biologically plausible than other actual maximization algorithms or deep learning strategy using back propagation.

8.6   Auto-CM and Spin Net as a Dynamic Artificial Auto Associative Memory

We have shown how Spin Net algorithm can use the Auto-CM weights in order to build up a complex Auto Associative Memory. The interaction between Auto-CM and Spin Net may also be dynamic. Spin Net may be activated during the learning phase of Auto-CM in an asynchronous parallelism. The exchange between these two systems shows that learning and recall processes are and have to be two sides of the same coin.

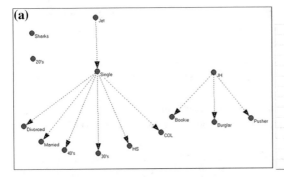

Inhibitions [0,-1]			
Cause		Effect	Strenght
Jet	->	Single	-0.16038
JH	->	Pusher	-0.11834
JH	->	Bookie	-0.11522
JH	->	Burglar	-0.10968
Single	->	30's	-0.09947
Single	->	40's	-0.10302
Single	->	COL	-0.10048
Single	->	HS	-0.10048
Single	->	Married	-0.10248
Single	->	Divorced	-0.10549

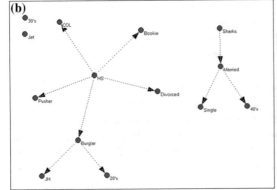

Inhibitions [0,-1]			
Cause		Effect	Strenght
Sharks	->	Married	-0.1421
HS	->	COL	-0.11967
HS	->	Divorced	-0.1142
HS	->	Pusher	-0.11359
HS	->	Bookie	-0.1159
HS	->	Burglar	-0.12278
Married	->	40's	-0.10846
Married	->	Single	-0.10738
Burglar	->	20's	-0.09908
Burglar	->	JH	-0.09913

Fig. 8.13 **a** Cause-effect inhibitory graph from the "Jets" perspective. **b** Cause-effect inhibitory graph from the "Sharks" perspective

The Spin Net algorithm possesses, from a biological point of view, many:

a. Spin Net uses only weights with positive values; in fact, the weights matrix generated by Auto-CM defines only the different strength of association among the variables;

b. The Spin Net cost function tries to minimize the energy difference between the external input, which works as the perception of the network from the environmental stimulus, and the energy already active into the network. The recursive application of this logic generates a deep learning process, where the network tries to read continuously its own interpretation of its perceptions;

c. Any external input represents the variables on which we intend to focus the attention of the ANN. The final output of the ANN represents the answer (pattern retrieval) through which the ANN has completed the external input.

d. The ANN dynamics from the input (external stimulus) up to the final restored pattern represents the specific job through which the ANN tries to give a meaning to the external input. This process shows the hidden rules that the ANN has used in order to build its memory (see TDM algorithm given in Sect. 8.5). Thus, if the external input is a type of "question" that it is posed to the ANN and if the final output represents the ANN "answer" to that question, the process connecting these two extreme points shows how the ANN is building up its working memory.

The last point is particularly interesting since it opens up the possibility to pose "questions" to a trained ANN. We can figure out at least three types of representative questions:

a. *Prototypical* questions;
b. *Outlier* questions;
c. *Impossible* questions;
d. *Virtual* questions.

We can prepare the external input for each of these questions in two different ways.

a. *Impulsive*: The external input is fed back into the ANN only at the first cycle of Spin Net. In this case we let the ANN (and the memory process) free to make its choice after the first cycle.
b. *Fixed*: The external input is fed back to the ANN at every new cycle of iteration. In this case the ANN reinforces and drives its memory process in specific directions.

8.7   Some Example of Prototypical Questions

We define a prototypical question when the external input focuses some variables that need to be complete in a reasonable way. In the Gang dataset we have provided the example of "Jets" prototype and of "Sharks" prototype. But any kind of binomial combination of variables may be set as the external input. For example we can look for the prototype of gang members that are "Jets" and "30's" simultaneously. And we can pose this question in two different ways: Impulsive and Fixed.

Figure 8.14 shows, according to the fixed mode, the final prototype of the variables "Jets" and "30's", and the patterns activation for this prototype. Spin Net associates to this prototype only with the variable "Single" and consequently makes more active only the patterns matching with this prototype. However, the patterns presenting only the variable "30's" are more active than the others. Figure 8.15 shows the dynamic through which Spin Net reaches the attractor of "Jets" and 30's.

Now we repeat the same question using the Impulsive mode. Figure 8.16 shows a completely different result: Spin Net took the decision to inhibit the "Jets" variable and emphasizes the "Sharks" variable and consequently all the variables and the pattern prototypes are changed. Figure 8.17 show this new dynamic is much more complex than the previous one.

Name	Gang	Age	Education	Status	Profession	Spin Net
ART	Jets	40age	JuniorScho	Single	Pusher	0.251564
AL	Jets	30age	JuniorScho	Married	Burglar	0.242968
SAM	Jets	20age	College	Single	Bookie	0.2513
CLYDE	Jets	40age	JuniorScho	Single	Bookie	0.2516
MIKE	Jets	30age	JuniorScho	Single	Bookie	0.697615
JIM	Jets	20age	JuniorScho	Divorced	Burglar	0.044681
GREG	Jets	20age	HighSchoo	Married	Pusher	0.0447
JOHN	Jets	20age	JuniorScho	Married	Burglar	0.044683
DOUG	Jets	30age	HighSchoo	Single	Bookie	0.697381
LANCE	Jets	20age	JuniorScho	Married	Burglar	0.044683
GEORGE	Jets	20age	JuniorScho	Divorced	Burglar	0.044681
PETE	Jets	20age	HighSchoo	Single	Bookie	0.251409
FRED	Jets	20age	HighSchoo	Single	Pusher	0.251373
GENE	Jets	20age	College	Single	Pusher	0.251264
RALPH	Jets	30age	JuniorScho	Single	Pusher	0.697574
PHIL	Sharks	30age	College	Married	Pusher	0.044281
IKE	Sharks	30age	JuniorScho	Single	Bookie	0.249873
NICK	Sharks	30age	HighSchoo	Single	Pusher	0.249629
DON	Sharks	30age	College	Married	Burglar	0.044217
NED	Sharks	30age	College	Married	Bookie	0.044289
KARL	Sharks	40age	HighSchoo	Married	Bookie	0.006712
KEN	Sharks	20age	HighSchoo	Single	Burglar	0.046174
EARL	Sharks	40age	HighSchoo	Married	Burglar	0.0067
RICK	Sharks	30age	HighSchoo	Divorced	Burglar	0.044239
OL	Sharks	30age	College	Married	Pusher	0.044281
NEAL	Sharks	30age	HighSchoo	Single	Bookie	0.249666
DAVE	Sharks	30age	HighSchoo	Divorced	Pusher	0.044303

Variables	Spin Net
Jet** =1.000	0.967663
Sharks	0.000036
20's	0.000048
30's** =1.000	0.963033
40's	0.000002
JH	0.000971
COL	0.000127
HS	0.000416
Single	0.985394
Married	0.000025
Divorced	0
Pusher	0.000826
Bookie	0.000923
Burglar	0.00007

Fig. 8.14 The prototype for "Jets" and "30's" variables with fixed external input

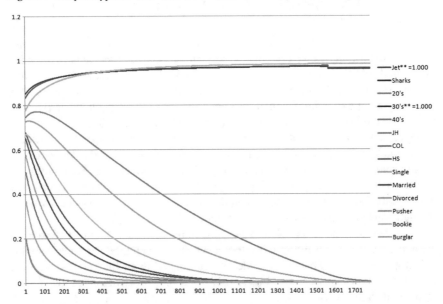

Fig. 8.15 The dynamics of Spin Net for "Jets" and "30's" variables with fixed external input after 1800 cycles

Variables	Spin Net
Jet** =1.000	0.000592
Sharks	0.949055
20's	0.00004
30's** =1.000	0.956099
40's	0
JH	0.230714
COL	0.297352
HS	0.334167
Single	0.354749
Married	0.365518
Divorced	0.117281
Pusher	0.461011
Bookie	0.355051
Burglar	0.156668

Name	Gang	Age	Education	Status	Profession	Spin Net
ART	Jets	40age	JuniorScho	Single	Pusher	0.051864
AL	Jets	30age	JuniorScho	Married	Burglar	0.17068
SAM	Jets	20age	College	Single	Bookie	0.048134
CLYDE	Jets	40age	JuniorScho	Single	Bookie	0.04238
MIKE	Jets	30age	JuniorScho	Single	Bookie	0.230483
JIM	Jets	20age	JuniorScho	Divorced	Burglar	0.018174
GREG	Jets	20age	HighSchoo	Married	Pusher	0.064324
JOHN	Jets	20age	JuniorScho	Married	Burglar	0.029514
DOUG	Jets	30age	HighSchoo	Single	Bookie	0.269201
LANCE	Jets	20age	JuniorScho	Married	Burglar	0.029514
GEORGE	Jets	20age	JuniorScho	Divorced	Burglar	0.018174
PETE	Jets	20age	HighSchoo	Single	Bookie	0.051622
FRED	Jets	20age	HighSchoo	Single	Pusher	0.063039
GENE	Jets	20age	College	Single	Pusher	0.058828
RALPH	Jets	30age	JuniorScho	Single	Pusher	0.270189
PHIL	Sharks	30age	College	Married	Pusher	0.742322
IKE	Sharks	30age	JuniorScho	Single	Bookie	0.666265
NICK	Sharks	30age	HighSchoo	Single	Pusher	0.75216
DON	Sharks	30age	College	Married	Burglar	0.610493
NED	Sharks	30age	College	Married	Bookie	0.699759
KARL	Sharks	40age	HighSchoo	Married	Bookie	0.270435
KEN	Sharks	20age	HighSchoo	Single	Burglar	0.196133
EARL	Sharks	40age	HighSchoo	Married	Burglar	0.199538
RICK	Sharks	30age	HighSchoo	Divorced	Burglar	0.506634
OL	Sharks	30age	College	Married	Pusher	0.742322
NEAL	Sharks	30age	HighSchoo	Single	Bookie	0.710588
DAVE	Sharks	30age	HighSchoo	Divorced	Pusher	0.653674

Fig. 8.16 The prototype for "Jets" and "30's" variables with impulsive external input

8.8   Some Example of Outlier Questions

Outlier questions are represented by an external input that makes active those variables that are weakly connected in the assigned dataset. In Gang dataset, for example, the variable "20's" has only a linear correlation of −0.5316 with the variable "Sharks". Thus, to force the ANN to combine these two variables in one prototype could be hard, because they present a strong inverse correlation.

Figure 8.18 shows the results of this prototype with external input fixed. Spin Net found the variable "Burglar" as a possible link between these two variables. Patterns presenting the variables "Burglar" are activated, especially "Ken" (the only Sharks member with the variable "20's" on). Jets members are activated only if they have on the variables "20's" and "Burglar" simultaneously. Figure 8.19 shows the dynamics of the process. It is interesting to observe how the landscape of the interaction among the variables changes suddenly at a specific cycle of the process. Around cycle # 300 "HS" starts a fast decrease and "Burglar" starts a fast increase. This is a typical of the highly non- linear effect of the Spin Net algorithm and Auto-CM weights.

Figures 8.20 and 8.21 show the behaviour of the networks when the same external input is run in impulsive mode. Spin Net rearranges the values of the variables along its dynamics and at the end it proposes a weak prototype of the "Sharks" member. It is important to outline that only two members of "Jets" are partially activated:

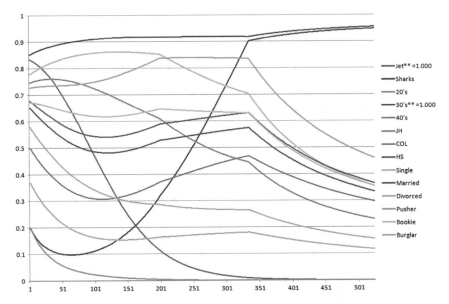

Fig. 8.17 The dynamics of Spin Net for "Jets" and "30's" variables with impulsive external input after 550 cycles

"Greg" and "Doug" who represent a conceptual bridge between the two gangs. In this case Spin Net takes the decision to "correct" the question itself.

8.9   Some Example of Impossible Questions

A question is "impossible" when the external input forces the networks to make active a set of variables whose reciprocal activation is not allowed in the given dataset. In Gang dataset, for example, a gang member may belong to the "Jets" or to the "Sharks", but not both of them. The same rule works for the macro variables "Age", "Education", "Status" and "Profession". If we formulate an impossible question (Query) in a relational database, it will generate a "NULL" answer. What happens when we simultaneously set active the variables "Jets" and "Sharks" in the gang dataset and we force the networks to process this impossible question with a fixed external input? This question makes sense: We desire to know which prototype is able to put together the features of both gangs. We also desire to know those members of the entire dataset that could belong to this meta-cluster.

Figure 8.22 shows the prototype of a hybrid member of both gangs where, in this case, only the variable "30's" can work as bridge and determine the gang members that could belong to this new prototype (in red colour). Figure 8.23 shows the complex dynamics of the algorithm associated with obtaining this solution.

Name	Gang	Age	Education	Status	Profession	Spin Net
ART	Jets	40age	JuniorSchool	Single	Pusher	0.006697
AL	Jets	30age	JuniorSchool	Married	Burglar	0.045937
SAM	Jets	20age	College	Single	Bookie	0.040524
CLYDE	Jets	40age	JuniorSchool	Single	Bookie	0.006697
MIKE	Jets	30age	JuniorSchool	Single	Bookie	0.006697
JIM	Jets	20age	JuniorSchool	Divorced	Burglar	0.231518
GREG	Jets	20age	HighSchool	Married	Pusher	0.040615
JOHN	Jets	20age	JuniorSchool	Married	Burglar	0.231737
DOUG	Jets	30age	HighSchool	Single	Bookie	0.006703
LANCE	Jets	20age	JuniorSchool	Married	Burglar	0.231737
GEORGE	Jets	20age	JuniorSchool	Divorced	Burglar	0.231518
PETE	Jets	20age	HighSchool	Single	Bookie	0.040558
FRED	Jets	20age	HighSchool	Single	Pusher	0.040559
GENE	Jets	20age	College	Single	Pusher	0.040525
RALPH	Jets	30age	JuniorSchool	Single	Pusher	0.006697
PHIL	Sharks	30age	College	Married	Pusher	0.039791
IKE	Sharks	30age	JuniorSchool	Single	Bookie	0.039738
NICK	Sharks	30age	HighSchool	Single	Pusher	0.039769
DON	Sharks	30age	College	Married	Burglar	0.228096
NED	Sharks	30age	College	Married	Bookie	0.03979
KARL	Sharks	40age	HighSchool	Married	Bookie	0.03982
KEN	Sharks	20age	HighSchool	Single	Burglar	0.649135
EARL	Sharks	40age	HighSchool	Married	Burglar	0.228235
RICK	Sharks	30age	HighSchool	Divorced	Burglar	0.228031
OL	Sharks	30age	College	Married	Pusher	0.039791
NEAL	Sharks	30age	HighSchool	Single	Bookie	0.039768
DAVE	Sharks	30age	HighSchool	Divorced	Pusher	0.039777

Variables	Spin Net
Jet	0.000048
Sharks** =1.000	0.907263
20's** =1.000	0.917498
30's	0.000037
40's	0
JH	0.000132
COL	0.000096
HS	0.000526
Single	0.0001
Married	0.000825
Divorced	0.00021
Pusher	0.000026
Bookie	0.000015
Burglar	0.982234

Fig. 8.18 The prototype for "Sharks" and "20's" variables with fixed external input

We observe a completely different dynamic if we feed the network with the same external input and with impulsive mode. The algorithm may choose to provide along its process the most harmonic solution from its point of view. Figure 8.24 shows how the network changes its original input and rearranges the values of its variables in order to define a weak prototype of the "Jets" member, whose dominant variable is to be "Single". Also the patterns selection for this prototype focuses only on those "Jets" members marked by the variable "Single". Figure 8.25 shows the sudden change of dynamics of the interaction of the variables when the network decides to rearrange the selection of pertinent features of the prototype.

8.10   Some Examples of Virtual Questions

A virtual question is defined by an external input whose combination of variables is not present in any pattern of the assigned dataset, but that combination is acceptable according to the logic of the dataset itself. In Gang dataset, for example, the combination of the variables "40's" and "College" is never present in the dataset. However,

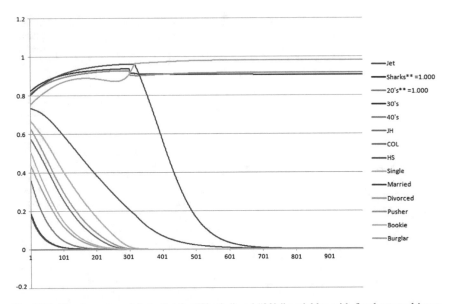

Fig. 8.19 The dynamics of Spin Net for "Sharks" and "20's" variables with fixed external input after 1000 cycles

Variables	Spin Net
Jet	0
Sharks** =1.000	0.968783
20's** =1.000	0.000142
30's	0.533555
40's	0.077377
JH	0.019438
COL	0.179119
HS	0.947392
Single	0.115026
Married	0.55582
Divorced	0.07198
Pusher	0.169893
Bookie	0.167373
Burglar	0.275918

Name	Gang	Age	Education	Status	Profession	Spin Net
ART	Jets	40age	JuniorSchool	Single	Pusher	0.014252
AL	Jets	30age	JuniorSchool	Married	Burglar	0.097041
SAM	Jets	20age	College	Single	Bookie	0.016681
CLYDE	Jets	40age	JuniorSchool	Single	Bookie	0.014181
MIKE	Jets	30age	JuniorSchool	Single	Bookie	0.034582
JIM	Jets	20age	JuniorSchool	Divorced	Burglar	0.013857
GREG	Jets	20age	HighSchool	Married	Pusher	0.160636
JOHN	Jets	20age	JuniorSchool	Married	Burglar	0.035662
DOUG	Jets	30age	HighSchool	Single	Bookie	0.18644
LANCE	Jets	20age	JuniorSchool	Married	Burglar	0.035662
GEORGE	Jets	20age	JuniorSchool	Divorced	Burglar	0.013857
PETE	Jets	20age	HighSchool	Single	Bookie	0.073092
FRED	Jets	20age	HighSchool	Single	Pusher	0.073434
GENE	Jets	20age	College	Single	Pusher	0.016764
RALPH	Jets	30age	JuniorSchool	Single	Pusher	0.034751
PHIL	Sharks	30age	College	Married	Pusher	0.453718
IKE	Sharks	30age	JuniorSchool	Single	Bookie	0.199144
NICK	Sharks	30age	HighSchool	Single	Pusher	0.615218
DON	Sharks	30age	College	Married	Burglar	0.506597
NED	Sharks	30age	College	Married	Bookie	0.452468
KARL	Sharks	40age	HighSchool	Married	Bookie	0.606706
KEN	Sharks	20age	HighSchool	Single	Burglar	0.404806
EARL	Sharks	40age	HighSchool	Married	Burglar	0.657141
RICK	Sharks	30age	HighSchool	Divorced	Burglar	0.64457
OL	Sharks	30age	College	Married	Pusher	0.453718
NEAL	Sharks	30age	HighSchool	Single	Bookie	0.614024
DAVE	Sharks	30age	HighSchool	Divorced	Pusher	0.594646

Fig. 8.20 The prototype for "Sharks" and "20's" variables with impulsive external input

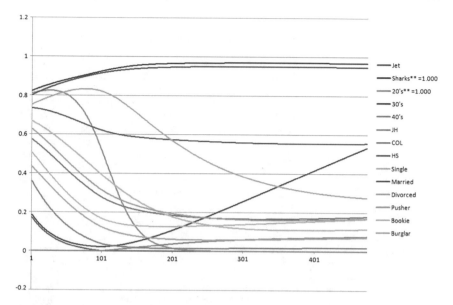

Fig. 8.21 The dynamics of Spin Net for "Sharks" and "20's" variables with impulsive external input after 500 cycles

this is theoretically possible. The problems we can resolve through these types of questions are many:

a. Which other variables could belong to this combination and which patterns are similar to it, when we force the external input;
b. Which kind of prototype will be proposed when we leave the network free to take its decision, after the first impulse, and which members of the Gang will be candidate for this new prototype.

Figure 8.26 shows the behavior of Spin Net when we fix the external input with the variables "40's" and "College". Only the variable "Married" is selected as common background of this combination, and only four "Sharks" members are candidate for this virtual prototype. Figure 8.27 shows the evolution of the network into the space of the phases.

The evolution of the network changes dramatically when we repeat the same question with an impulsive input,. The variable "College" remains active, but the variable "40's" decreases and "30's" takes its place. The final prototype includes the most variables representing the "Sharks", but also one variable representing the "Jets" ("Pusher"). Some of the patterns selected, in fact, are also "Jets" members (see Figs. 8.28 and 8.29).

Name	Gang	Age	Education	Status	Profession	Spin Net
ART	Jets	40age	JuniorSchool	Single	Pusher	0.035434
AL	Jets	30age	JuniorSchool	Married	Burglar	0.207193
SAM	Jets	20age	College	Single	Bookie	0.035424
CLYDE	Jets	40age	JuniorSchool	Single	Bookie	0.035416
MIKE	Jets	30age	JuniorSchool	Single	Bookie	0.207271
JIM	Jets	20age	JuniorSchool	Divorced	Burglar	0.035342
GREG	Jets	20age	HighSchool	Married	Pusher	0.035485
JOHN	Jets	20age	JuniorSchool	Married	Burglar	0.0354
DOUG	Jets	30age	HighSchool	Single	Bookie	0.207494
LANCE	Jets	20age	JuniorSchool	Married	Burglar	0.0354
GEORGE	Jets	20age	JuniorSchool	Divorced	Burglar	0.035342
PETE	Jets	20age	HighSchool	Single	Bookie	0.035462
FRED	Jets	20age	HighSchool	Single	Pusher	0.03548
GENE	Jets	20age	College	Single	Pusher	0.035441
RALPH	Jets	30age	JuniorSchool	Single	Pusher	0.207355
PHIL	Sharks	30age	College	Married	Pusher	0.218265
IKE	Sharks	30age	JuniorSchool	Single	Bookie	0.218116
NICK	Sharks	30age	HighSchool	Single	Pusher	0.218435
DON	Sharks	30age	College	Married	Burglar	0.218073
NED	Sharks	30age	College	Married	Bookie	0.218178
KARL	Sharks	40age	HighSchool	Married	Bookie	0.037751
KEN	Sharks	20age	HighSchool	Single	Burglar	0.037724
EARL	Sharks	40age	HighSchool	Married	Burglar	0.037729
RICK	Sharks	30age	HighSchool	Divorced	Burglar	0.217976
OL	Sharks	30age	College	Married	Pusher	0.218265
NEAL	Sharks	30age	HighSchool	Single	Bookie	0.218348
DAVE	Sharks	30age	HighSchool	Divorced	Pusher	0.218168

Variables	Spin Net
Jet** =1.000	0.846006
Sharks** =1.000	0.878393
20's	0
30's	0.981541
40's	0.000003
JH	0.000146
COL	0.000257
HS	0.000824
Single	0.000884
Married	0.000954
Divorced	0.000102
Pusher	0.000952
Bookie	0.000697
Burglar	0.00039

Fig. 8.22 The prototype for "Sharks" and "Jets" variables with fixed external input

8.11 Content Addressable Memory and Highly Nonlinear Systems

We have shown as Spin Net with the weights matrix generated by Auto-CM is able to behave as a good dynamic auto-associative memory. While a relational data base works as a tautological system, Spin Net shows the specific behaviour of a Content Addressable Memory, able to make analogies among the contents and able to provide new abstract contents.

Spin net also shows a highly nonlinear behaviour. In other words: if we increase linearly the external input, the dynamics of the process starts to jump suddenly from one state to another one. This is the typical behaviour of a multi-phase system. This high nonlinearity is also present when the external input is impulsive (that is, without any forcing from external). We will provide an example of this highly nonlinearity, using again the Gang dataset.

Let us imagine that the value of three external input variables to 0.1. Then we let the network manifests its dynamics until it reaches the attractor. At this point, we increase the values of the three variables of the external input by 0.01, and so on, until

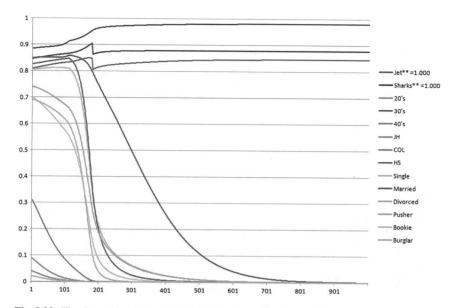

Fig. 8.23 The dynamics of Spin Net for "Sharks" and "Jets" variables with fixed external input after 1000 cycles

Variables	Spin Net
Jet** =1.000	0.979536
Sharks** =1.000	0.000084
20's	0.690813
30's	0.050521
40's	0.007028
JH	0.685982
COL	0.000679
HS	0.057535
Single	0.962821
Married	0.07651
Divorced	0
Pusher	0.196342
Bookie	0.147868
Burglar	0.112327

Name	Gang	Age	Education	Status	Profession	Spin Net
ART	Jets	40age	JuniorSchool	Single	Pusher	0.660028
AL	Jets	30age	JuniorSchool	Married	Burglar	0.233215
SAM	Jets	20age	College	Single	Bookie	0.637246
CLYDE	Jets	40age	JuniorSchool	Single	Bookie	0.637948
MIKE	Jets	30age	JuniorSchool	Single	Bookie	0.657789
JIM	Jets	20age	JuniorSchool	Divorced	Burglar	0.484334
GREG	Jets	20age	HighSchool	Married	Pusher	0.269231
JOHN	Jets	20age	JuniorSchool	Married	Burglar	0.522569
DOUG	Jets	30age	HighSchool	Single	Bookie	0.353557
LANCE	Jets	20age	JuniorSchool	Married	Burglar	0.522569
GEORGE	Jets	20age	JuniorSchool	Divorced	Burglar	0.484334
PETE	Jets	20age	HighSchool	Single	Bookie	0.663101
FRED	Jets	20age	HighSchool	Single	Pusher	0.684406
GENE	Jets	20age	College	Single	Pusher	0.659346
RALPH	Jets	30age	JuniorSchool	Single	Pusher	0.679267
PHIL	Sharks	30age	College	Married	Pusher	0.012721
IKE	Sharks	30age	JuniorSchool	Single	Bookie	0.21325
NICK	Sharks	30age	HighSchool	Single	Pusher	0.07832
DON	Sharks	30age	College	Married	Burglar	0.010774
NED	Sharks	30age	College	Married	Bookie	0.011559
KARL	Sharks	40age	HighSchool	Married	Bookie	0.011868
KEN	Sharks	20age	HighSchool	Single	Burglar	0.205406
EARL	Sharks	40age	HighSchool	Married	Burglar	0.011063
RICK	Sharks	30age	HighSchool	Divorced	Burglar	0.010363
OL	Sharks	30age	College	Married	Pusher	0.012721
NEAL	Sharks	30age	HighSchool	Single	Bookie	0.071602
DAVE	Sharks	30age	HighSchool	Divorced	Pusher	0.012236

Fig. 8.24 The prototype for "Sharks" and "Jets" variables with impulsive external input

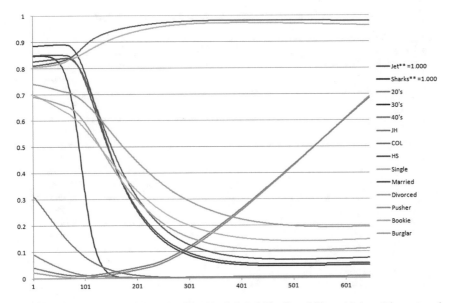

Fig. 8.25 The dynamics of Spin Net for "Sharks" and "Jets" variables with impulsive external input after 650 cycles

the maximum value allowed for the external input (that is 1.0). With this approach, we can generate 100 meta-cycles, each one composed of a specific number of cycles. For each meta-cycle we have a specific attractor, thus we can plot the 100 attractors all together, each one with the specific value of each variable of the dataset (y axis), in relation to the linear increase of the external input of the three variables selected (x axis).

We have decided to use the variables "Pusher", "Bookie" and "Burglar" as control variables in this example. Figure 8.30 shows the 100 Meta cycles of this complex query.

A synthetic index of a strong change in a dynamical system is the entropy differ- ence between one state and the other one. We have calculated the Shannon entropy of each meta-cycle according to the following equation:

$$S^{[t]} = -\sum_i^N \frac{u_i^{[n]}}{\sum_k^N u_i^{[n]}} \cdot \log_2 \left(\frac{u_i^{[n]}}{\sum_k^N u_i^{[n]}} \right) \tag{8.22}$$

where:

$u_i^{[t]}$ = Final activation of the i-th unit at [t] metacycle;
$u_i^{[n]}$ = Final activation of the i-th unit at [n] metacycle;
N = Number of the variables.

Variables	Spin Net
Jet	0.000218
Sharks	0.001014
20's	0.000108
30's	0.000172
40's** =1.000	0.857596
JH	0.000105
COL** =1.000	0.877394
HS	0.000118
Single	0.00007
Married	0.964281
Divorced	0
Pusher	0.000394
Bookie	0.000417
Burglar	0.000246

Name	Gang	Age	Education	Status	Profession	Spin Net
ART	Jets	40age	JuniorSchool	Single	Pusher	0.036151
AL	Jets	30age	JuniorSchool	Married	Burglar	0.044364
SAM	Jets	20age	College	Single	Bookie	0.037558
CLYDE	Jets	40age	JuniorSchool	Single	Bookie	0.036153
MIKE	Jets	30age	JuniorSchool	Single	Bookie	0.006706
JIM	Jets	20age	JuniorSchool	Divorced	Burglar	0.006702
GREG	Jets	20age	HighSchool	Married	Pusher	0.044372
JOHN	Jets	20age	JuniorSchool	Married	Burglar	0.044358
DOUG	Jets	30age	HighSchool	Single	Bookie	0.006706
LANCE	Jets	20age	JuniorSchool	Married	Burglar	0.044358
GEORGE	Jets	20age	JuniorSchool	Divorced	Burglar	0.006702
PETE	Jets	20age	HighSchool	Single	Bookie	0.006705
FRED	Jets	20age	HighSchool	Single	Pusher	0.006705
GENE	Jets	20age	College	Single	Pusher	0.037556
RALPH	Jets	30age	JuniorSchool	Single	Pusher	0.006706
PHIL	Sharks	30age	College	Married	Pusher	0.211903
IKE	Sharks	30age	JuniorSchool	Single	Bookie	0.006717
NICK	Sharks	30age	HighSchool	Single	Pusher	0.006716
DON	Sharks	30age	College	Married	Burglar	0.211854
NED	Sharks	30age	College	Married	Bookie	0.211911
KARL	Sharks	40age	HighSchool	Married	Bookie	0.205356
KEN	Sharks	20age	HighSchool	Single	Burglar	0.006714
EARL	Sharks	40age	HighSchool	Married	Burglar	0.2053
RICK	Sharks	30age	HighSchool	Divorced	Burglar	0.006713
OL	Sharks	30age	College	Married	Pusher	0.211903
NEAL	Sharks	30age	HighSchool	Single	Bookie	0.006717
DAVE	Sharks	30age	HighSchool	Divorced	Pusher	0.006715

Fig. 8.26 The prototype for "40's" and "College" variables with fixed external input

A first sharp transformation occurs between meta cycle 10 and meta cycle 11. Thus, when the values of the three control variables change from 0.1 to 0.11, the process changes its state suddenly and nonlinearly. Another change of this type is between meta-cycle 19 and meta-cycle 20 (control variables pass from 0.19 to 0.20). Figure 8.31 shows the dynamics of the Entropy in these critic points. In Fig. 8.32a, b these key transformations are compared to the final meta-cycle, when all the three external input are at 1.00.

This short example shows the dynamic memory simulated by Spin Net and Auto-CM is a powerful engine able to generate an exponential number of highly non-linear processes, whose structure is very similar to the structure of what we know about human memory. Obviously we know we are only at the beginning of our investigations.

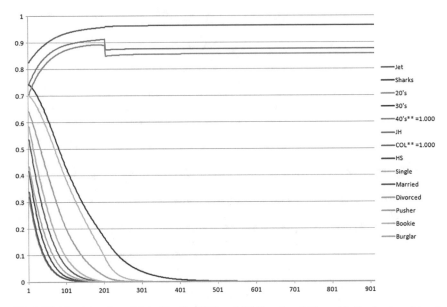

Fig. 8.27 The dynamics of Spin Net for "40" and "College" variables with fixed external input after 910 cycles

8.12   Discussion: Deep Learning Versus Fat Learning

Marvin Minsky, in 1988, wrote that the brain is not simulated by single big neural networks. According to Minsky, in fact, the human brain is similar to a complex society of neural networks, each one working in a specific way and with a specific topology [14].

The actual promoters of the Deep Learning [15–18] strategy do not consider Minsky's suggestion. They prefer to pursue the technological target of effectiveness; a single neural network with enough layers of hidden units to be able to emulate many human abilities (as example human sight). We do not dispute the fact that you can teach a frog to ride a bike. We also observe that it is useless to build a frog who writes like Shakespeare.

The Multilayer Perceptron is less complex than a frog in relation to human brain. And this for many reasons:

a.   The back propagation of signals from output to input has no biological evidence;
b.   Synapses do not change their excitatory or inhibitory nature according to the learning task;
c.   We conjecture that supervised ANNs are mathematically and ontogenetically a specific case of the unsupervised ANNs. Thus targets in supervised learning must not be imposed a priori from the researcher, but they should emerge spontaneously from unsupervised learning. We leave it to the reader as an open question;

Name	Gang	Age	Education	Status	Profession	Spin Net
ART	Jets	40age	JuniorSchool	Single	Pusher	0.026842
AL	Jets	30age	JuniorSchool	Married	Burglar	0.121183
SAM	Jets	20age	College	Single	Bookie	0.151153
CLYDE	Jets	40age	JuniorSchool	Single	Bookie	0.01829
MIKE	Jets	30age	JuniorSchool	Single	Bookie	0.037777
JIM	Jets	20age	JuniorSchool	Divorced	Burglar	0.013921
GREG	Jets	20age	HighSchool	Married	Pusher	0.168312
JOHN	Jets	20age	JuniorSchool	Married	Burglar	0.085718
DOUG	Jets	30age	HighSchool	Single	Bookie	0.037781
LANCE	Jets	20age	JuniorSchool	Married	Burglar	0.085718
GEORGE	Jets	20age	JuniorSchool	Divorced	Burglar	0.013921
PETE	Jets	20age	HighSchool	Single	Bookie	0.026002
FRED	Jets	20age	HighSchool	Single	Pusher	0.03802
GENE	Jets	20age	College	Single	Pusher	0.208625
RALPH	Jets	30age	JuniorSchool	Single	Pusher	0.054931
PHIL	Sharks	30age	College	Married	Pusher	0.827261
IKE	Sharks	30age	JuniorSchool	Single	Bookie	0.086511
NICK	Sharks	30age	HighSchool	Single	Pusher	0.122977
DON	Sharks	30age	College	Married	Burglar	0.689335
NED	Sharks	30age	College	Married	Bookie	0.763865
KARL	Sharks	40age	HighSchool	Married	Bookie	0.187088
KEN	Sharks	20age	HighSchool	Single	Burglar	0.042303
EARL	Sharks	40age	HighSchool	Married	Burglar	0.136341
RICK	Sharks	30age	HighSchool	Divorced	Burglar	0.047704
OL	Sharks	30age	College	Married	Pusher	0.827261
NEAL	Sharks	30age	HighSchool	Single	Bookie	0.086519
DAVE	Sharks	30age	HighSchool	Divorced	Pusher	0.097568

Variables	Spin Net
Jet	0.10703
Sharks	0.547295
20's	0.179897
30's	0.372798
40's** =1.000	0.000102
JH	0.000141
COL** =1.000	0.949018
HS	0.000198
Single	0.130001
Married	0.94662
Divorced	0
Pusher	0.467438
Bookie	0.271262
Burglar	0.082774

Fig. 8.28 The prototype for "40's" and "College" variables with impulsive external input

d. The recall of a trained ANN has to be a dynamic process. It cannot be produced by a one shot output.

In other words, we think the effectiveness of the classic deep learning strategy is based more on brute force, than on a deep understanding of how the brain works. From our point of view it is a "fat learning", because it is similar to a food with many saturated fats. From our perspective, effective deep learning has to be based on specific requirements:

a. The initial weights of every ANN may not be randomly chosen;
b. The updating of weights has to occur during the signal transfer and not by the back propagation of the signal error;
c. Each learning task has to be executed from different and specialized ANNs connected to each other by a specific topology;
d. Each ANN has to be able to work also as an Artificial Auto Associative Memory: learning and dynamic recall are two sides of the same coin;
e. Each Artificial Auto Associative Memory needs to work in recursive mode, in order to minimize the difference between the perception (representation) it has of the external input and its internal states;

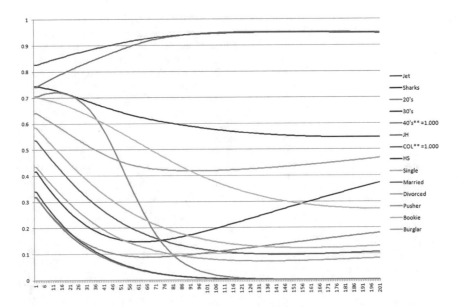

Fig. 8.29 The dynamics of Spin Net for "40" and "College" variables with impulsive external input after 202 cycles

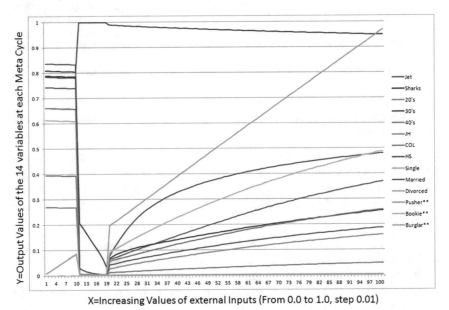

Fig. 8.30 The dynamics of 100 attractors when the control variables (**) are increased linearly

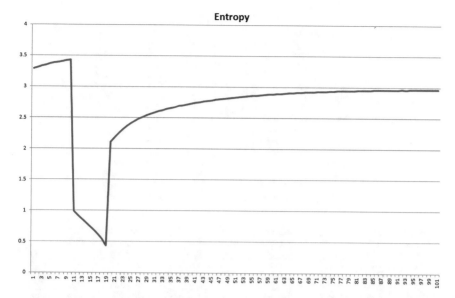

Fig. 8.31 The dynamics of Shannon Entropy when the control variables are increased linearly

f. The temporal process through which an Artificial Auto Associative Memory manages every external input has to provide new information about the past learning of this ANN and the novelties of the stimulations coming from external world.

We think that Spin Net and Auto-CM satisfy these requirements.

(a)

| Meta Cycle= | 10 | | Meta Cycle= | 11 | |
| Entropy= | 3.429858 | | Entropy= | 0.991829 | |
Attributes	Input	Output	Attributes	Input	Output
Jet	0.00	0.8329	Jet	0.00	0.0012
Sharks	0.00	0.7846	Sharks	0.00	0.2068
20's	0.00	0.6583	20's	0.00	0.0000
30's	0.00	0.8047	30's	0.00	0.9981
40's	0.00	0.2678	40's	0.00	0.0000
JH	0.00	0.6572	JH	0.00	0.0019
COL	0.00	0.3931	COL	0.00	0.0036
HS	0.00	0.7806	HS	0.00	0.0297
Single	0.00	0.6092	Single	0.00	0.0048
Married	0.00	0.7394	Married	0.00	0.0076
Divorced	0.00	0.0000	Divorced	0.00	0.0017
Pusher	0.10	0.0867	Pusher	0.11	0.0039
Bookie	0.10	0.0860	Bookie	0.11	0.0005
Burglar	0.10	0.0847	Burglar	0.11	0.0003

(b)

| Meta Cycle= | 19 | | Meta Cycle= | 20 | | Meta Cycle= | 100 | |
| Entropy= | 0.430797 | | Entropy= | 2.103621 | | Entropy= | 2.966085 | |
Attributes	Input	Output	Attributes	Input	Output	Attributes	Input	Output
Jet	0.00	0.0012	Jet	0.00	0.0646	Jet	0.00	0.3697
Sharks	0.00	0.0343	Sharks	0.00	0.0678	Sharks	0.00	0.2547
20's	0.00	0.0000	20's	0.00	0.0275	20's	0.00	0.1600
30's	0.00	0.9989	30's	0.00	0.9901	30's	0.00	0.9489
40's	0.00	0.0000	40's	0.00	0.0000	40's	0.00	0.0000
JH	0.00	0.0019	JH	0.00	0.0127	JH	0.00	0.0483
COL	0.00	0.0028	COL	0.00	0.0576	COL	0.00	0.2588
HS	0.00	0.0034	HS	0.00	0.0797	HS	0.00	0.4803
Single	0.00	0.0032	Single	0.00	0.0930	Single	0.00	0.4877
Married	0.00	0.0033	Married	0.00	0.0426	Married	0.00	0.1880
Divorced	0.00	0.0011	Divorced	0.00	0.0022	Divorced	0.00	0.0027
Pusher	0.19	0.0065	Pusher	0.20	0.1972	Pusher	1.00	0.9704
Bookie	0.19	0.0005	Bookie	0.20	0.0001	Bookie	1.00	0.0001
Burglar	0.19	0.0003	Burglar	0.20	0.0001	Burglar	1.00	0.0001

Fig. 8.32 a Transformation of the attractors generated by Spin Net in impulsive mode, when the three external inputs change their values from 0.10 to 0.11 (in red colour the relevant changes). **b** Transformations of the attractors generated by Spin Net in impulsive mode, when the three external inputs change their values from 0.19 to 0.20 and from 0.20 to 1.0 (in red colour the relevant changes)

References

1. Hawkins, J., and S. Blakeslee. 2004. On Intelligence: How a New Understanding of the Brain will lead to Truly Intelligent Machines. Henry Holt and Company, Jeff Hawkins, and Subutai Ahmad. 2016. Why Neurons Have Thousands of Synapses, A Theory of Sequence Memory in Neocortex, Front Neural Circuits, vol. 10, 23. Published online 30 March, 2016. https://doi.org/10.3389/fncir.2016.00023.
2. Associative ANNs, Semeion Software #51, version 30.0, 2008-2016, Semeion, Rome, Italy (see www.semeion.t).
3. Spin Net was designed by Massimo Buscema in 2013 at Semeion Research Center in Rome Italy. Spin Net is implemented in *Modular Associative ANNs Semeion Software #5 Version 30.0, 200, 2016 and in Path Net Semeion Software #58, Version 8.0, 2013, 2015* Semeion Research Center in Rome, Italy.
4. Grossberg, S. 1976. Adaptive Pattern Classification and Universal Recoding: Part I: Parallel Development and Coding of Neural Feature Detectors. *Biological Cybernetics* 23: 121–134. PMID: 974165.

5. Grossberg, S. 1978. A Theory of Visual Coding, Memory, and Development. In *Formal Theories of Visual Perception*, ed. E.L.J. Leeuwenberg, and H.F.J.M. Buffart. New York: Wiley.

6. Grossberg, S. 1980. How Does the Brain Build a Cognitive Code? *Psychological Review* 87: 1–51. PMID: 7375607.

7. McClelland, J., and D.E. Rumelhart. 1981. An Interactive Activation Model of Context Effects in Letter Perception: Part 1. An Account of Basic Findings. *Psychological Review* 88: 375–407.

8. McClelland, J., and D. Rumelhart. 1988. *Explorations in Parallel Distributed Processing: A Handbook of Models, Programs, and Exercises*. The MIT Press.

9. Tsang, E. 1993. *Foundations of Constraint Satisfaction*. Academic Press.

10. Ming Mao, Yefei Peng, and Michael Spring. 2008. *Neural Network Based Constraint Satisfaction in Ontology Mapping*. University of Pittsburgh, Association for the Advancement of Artificial Intelligence. www.aaai.org.

11. Mostafa, Hesham, Lorenz K. Müller, and Giacomo Indiveri. 2015. Rhythmic Inhibition Allows Neural Networks to Search for Maximally Consistent States. *Neural Computation* 27 (12): 2510–2547.

12. Braga, D. S., L. B. Gomide, J. S. S. Melo, M. T. D. Melo, and L. M. Brasil. 2015. Artificial Neural Network Interactive Activation and Competition Model Service-Oriented Applied to Health. In *IFMBE Proceedings of World Congress on Medical Physics and Biomedical Engineering*, June 7–12, 2015, Toronto, Canada, vol. 51, 1754–1757.

13. Target Diffusion Model (TDM) was designed by Massimo Buscema in 2012 at Semeion Research Center in Rome. TDM algorithm is implemented in *Massimo Buscema, P_FAST, Semeion Software #52, version 9.4, 2008-2013, Semeion, Rome, Italy* (see www.semeion.t).

14. Minsky, Marvin. 1988. *The Society of Mind*. New York: Simon & Schuster.

15. Bengio, Y. 2009. Learning Deep Architectures for AI. *Machine Learning* 2 (1): 1–127.

16. Raiko, T., H. Valpola, and Y. LeCun. 2012. Deep Learning Made Easier by Linear Transformations in Perceptrons. In *Proceedings of the 15th International Conference on Artificial Intelligence and Statistics*.

17. Hinton, G.E., and R.R. Salakhutdinov. 2006. Reducing the Dimensionality of Data with Neural Networks. *Science* 313 (5786): 504–507.

18. McCLelland, J. L., D. E. Rumelhart, G. E. Hinton. The Appeal of Parallel Distributed Processing, Chapter 1 (J. L. McClelland, and D. E. Rumelhart (eds.)), 3–44.

Printed in the United States
By Bookmasters